T0336712

Evolution and Standardization of Mobile Communications Technology

DongBack Seo
Hansung University, Republic of Korea

A volume in the Advances in IT Standards
and Standardization Research (AITSSR)
Book Series

An Imprint of IGI Global

Managing Director:	Lindsay Johnston
Editorial Director:	Joel Gamon
Production Manager:	Jennifer Yoder
Publishing Systems Analyst:	Adrienne Freeland
Development Editor:	Christine Smith
Assistant Acquisitions Editor:	Kayla Wolfe
Typesetter:	Lisandro Gonzalez
Cover Design:	Jason Mull

Published in the United States of America by
Information Science Reference (an imprint of IGI Global)
701 E. Chocolate Avenue
Hershey PA 17033
Tel: 717-533-8845
Fax: 717-533-8661
E-mail: cust@igi-global.com
Web site: http://www.igi-global.com

Library of Congress Cataloging-in-Publication Data

Seo, DongBack, 1971-
  Evolution and standardization of mobile communications technology / by DongBack Seo.
    pages cm
  Includes bibliographical references and index.
  Summary: "This book examines methods of developing and regulating compatibility standards in the ICT industry, assisting organizations in their application of the latest communications technologies in their business practices"--Provided by publisher.
  ISBN 978-1-4666-4074-0 (hardcover) -- ISBN 978-1-4666-4075-7 (ebook) -- ISBN 978-1-4666-4076-4 (print & perpetual access) 1. Mobile communication systems--Standards. I. Title.
  TK5103.2.S464 2013
  621.3845'60218--dc23
                              2013001586

This book is published in the IGI Global book series Advances in IT Standards and Standardization Research (AITSSR) Book Series (ISSN: 1935-3391; eISSN: 1935-3405)

British Cataloguing in Publication Data
A Cataloguing in Publication record for this book is available from the British Library.

All work contributed to this book is new, previously-unpublished material. The views expressed in this book are those of the authors, but not necessarily of the publisher.

Advances in IT Standards and Standardization Research (AITSSR) Book Series

Kai Jakobs
RWTH Aachen University, Germany

ISSN: 1935-3391
EISSN: 1935-3405

Mission

IT standards and standardization are a necessary part of effectively delivering IT and IT services to organizations and individuals as well as streamlining IT processes and minimizing organizational cost. In implementing IT standards, it is necessary to take into account not only the technical aspects, but also the characteristics of the specific environment where these standards will have to function.

The **Advances in IT Standards and Standardization Research (AITSSR) Book Series** seeks to advance the available literature on the use and value of IT standards and standardization. This research provides insight into the use of standards for the improvement of organizational processes and development in both private and public sectors.

Coverage

- Analyses of standards-setting processes, products, and organization
- Descriptive theory of standardization
- Emerging roles of formal standards organizations and consortia
- Intellectual property rights
- Management of standards
- National, regional, international and corporate standards strategies
- Open Source and standardization
- Risks of standardization
- Technological innovation and standardization
- User-related issues

IGI Global is currently accepting manuscripts for publication within this series. To submit a proposal for a volume in this series, please contact our Acquisition Editors at Acquisitions@igi-global.com or visit: http://www.igi-global.com/publish/.

Titles in this Series

For a list of additional titles in this series, please visit: www.igi-global.com

Evolution and Standardization of Mobile Communications Technology
DongBack Seo (University of Groningen, The Netherlands and Hansung University, South Korea)
Information Science Reference • copyright 2013 • 328pp • H/C (ISBN: 9781466640740) • US $195.00 (our price)

Information Technology for Intellectual Property Protection Interdisciplinary Advancements
Hideyasu Sasaki (Ritsumeikan University, Japan)
Information Science Reference • copyright 2012 • 367pp • H/C (ISBN: 9781613501351) • US $195.00 (our price)

Frameworks for ICT Policy Government, Social and Legal Issues
Esharenana E. Adomi (Delta State University, Nigeria)
Information Science Reference • copyright 2011 • 352pp • H/C (ISBN: 9781616920128) • US $180.00 (our price)

Toward Corporate IT Standardization Management Frameworks and Solutions
Robert van Wessel (Tilburg University, Netherlands)
Information Science Reference • copyright 2010 • 307pp • H/C (ISBN: 9781615207596) • US $180.00 (our price)

Data-Exchange Standards and International Organizations Adoption and Diffusion
Josephine Wapakabulo Thomas (Rolls-Royce, UK)
Information Science Reference • copyright 2010 • 337pp • H/C (ISBN: 9781605668321) • US $180.00 (our price)

Information Communication Technology Standardization for E-Business Sectors Integrating Supply and Demand Factors
Kai Jakobs (Aachen University, Germany)
Information Science Reference • copyright 2009 • 315pp • H/C (ISBN: 9781605663203) • US $195.00 (our price)

Standardization and Digital Enclosure The Privatization of Standards, Knowledge, and Policy in the Age of Global Information Technology
Timothy Schoechle (University of Colorado, USA)
Information Science Reference • copyright 2009 • 384pp • H/C (ISBN: 9781605663340) • US $165.00 (our price)

Standardization Research in Information Technology New Perspectives
Kai Jakobs (Aachen University, Germany)
Information Science Reference • copyright 2008 • 300pp • H/C (ISBN: 9781599045610) • US $180.00 (our price)

www.igi-global.com

701 E. Chocolate Ave., Hershey, PA 17033
Order online at www.igi-global.com or call 717-533-8845 x100
To place a standing order for titles released in this series, contact: cust@igi-global.com
Mon-Fri 8:00 am - 5:00 pm (est) or fax 24 hours a day 717-533-8661

Table of Contents

Foreword

by Mostafa Hashem Sherif

In this ambitious and pioneering book, Professor Seo offers an innovative approach to standardization using the paradigm of self-organization complexity. She provides a comprehensive framework—consistent with the dominant *zeitgeist*—to describe the gradual emergence of telecommunication standards in the absence of central coordination or control. She does this by relying on the basic premise of the Actor Network Theory (ANT), that is, the pursuit of self-interest, as the stimulus for self-organization.

In this approach, each entity is described by an organizational profile comprised of six elements: its capabilities, market position, the availability of complementary products, of alternative or substitute technologies, the nature of the innovation itself, and the characteristics of its intellectual property. Actions are constrained by the web of relationships within the emerging value network, the shaping of the standards-setting environment, and the willingness to share intellectual property. The evolution of the dominant configuration in a given industrial sector over time adds a dynamic dimension to the description.

Mobile technology standards are tracked over seven decades and three continents. The focus on Japan, Korea, and China in addition to North America and Western Europe is an important contribution of this book. This progress took place as a consequence of many technological innovations as well as changes to the financial and business environments and to the institutional set-up, thereby redefining the boundaries of possible actions and the meaning of self-interest. Even the terms by which the technology is denoted evolved over this period, starting as radio communication then becoming cellular, mobile, and wireless communication. Clearly, as Professor Seo indicates, the book is an introductory step to further refinements. Further investigation could bring forth the contribution of the network effect as the "invisible hand" that encourages all actors to compromise and reach a stable solution. The publication of this work could also encourage efforts to exploit newly published historical papers to improve the framework. Finally, it would be very helpful from a management of technology viewpoint to obtain a prescriptive method using the framework as a practical guide for strategy formulation and implementation in the field of standardization.

Mostafa Hashem Sherif
AT&T, USA

Mostafa Hashem Sherif *is a member of the Technical Staff at AT&T, a Senior Member of the IEEE, and the Standards Co-Editor of the IEEE Communications Magazine. He is also a (co)author of several books.*

Foreword

by Kai Jakobs

This is the first time I've been asked to write a foreword, and I feel honoured (and flattered) and indeed privileged to write a foreword for this highly interesting book.

In fact, in a way there are three books in one. First, we have a history book. While I suspect that this is rather more unintentional, the book does offer a nice account of the more recent developments in the cellular communication sector, from the various analogue 1G system of the 1960s and 1970s to the hugely popular digital 2G systems (most notably GSM) of the 1990s to the multi-media 3G systems most of us are using today. From a standards point of view, the book shows how the different generations emerged, why they emerged the way they did, and which stakeholders were driving the process.

The second book is about theory building and application. The "Self-Organized Complexity Unfolding Model" is a combination of Actor Network Theory and the theory of Self-Organised Complexity. It is shown to be very useful for the task at hand as it enables a holistic analysis of an organisation's standardisation strategy. Looking at one particular point in time of the "unfolding" process, the "Cross-Sectional Framework for Organizational Standards Strategy" provides us with a tool to analyse how a firm's standards strategy is shaped by a particular situation in time and the firm's (subjective) interpretation of this situation.

Thirdly, there is the management book. It offers descriptions of each individual stakeholder's situation, its perceptions and interpretations of this situation, the resulting actions and strategy, and the ultimate outcome of the interactions of the individual strategies. With the benefit of hindsight, we can now evaluate which of these strategies were successful and which were not. This, in turn, should inform future research activities and, hopefully, management and policy decisions.

I would consider especially the latter as crucially important. Research in fields like Management Studies should not just be *l'art pour l'art* (as far as one can see; this may sometimes be different in the Science and Engineering fields) but have rather immediate practical ramifications. That is, ideally this research should inform both scholars and—most notably, in my opinion—practitioners. DongBack's thorough analysis does exactly that.

Above, I have mentioned the "history" dimension of the book. Most historians I know are extremely reluctant when asked to draw some conclusions from their insights into historical developments that might be helpful in today's environment. They have a point. Lessons that may have been valid during, say, Charlemagne's times (I'm from Aachen) will hardly be applicable today. The respective environments are just too different. However, I, nonetheless, happen to believe that lessons can be learned from history. The technologies analysed in this book are from the 1960s onward. Since then, society, international relations, and, specifically, ICT have changed faster than in any other period in history. Yet, I believe that the boundary conditions are sufficiently similar today to allow us to draw conclusions from past experiences and events discussed in this book that will remain valid for a while. Moreover, the six elements of an organisational situation identified by DongBack that form the basis of her study are "time invariant"—they can be applied for an analysis of the current situation, for the (not too distant) past and for the (also not too distant) future.

The book also nicely highlights the importance of another trend that is still ongoing today—globalisation. The development of 1G systems was pretty much a national affair—individual states developed their individual national systems. For 2G, the scope of standardisation had moved to the "regional" level, with basically four developments (two of which survived) going on in parallel in three parts of the world. During this phase, we also saw trans-regional alliances in support of a technology. Thus, it is a logical extension of this trend that from 3G onward standardisation of cellular systems has become a global affair. While this may be seen as old news, managers, policy makers, as well as researchers need to be very aware of the fact that standardisation developments going on in, say, Korea, may well have significant ramifications for Europe and European firms, and vice versa. ICT is a "global" technology, and both companies and nation states need to take this into account.

The book highlights two other—interlinked—lessons. Number one is that even robust-looking trajectories may well be interrupted and number two would be that such an interruption may well be caused by one single stubborn firm with a superior technology. Obviously, this makes strategists' and policy makers' lives more complicated. On the other hand, it clearly demonstrates the need for at least manufacturers and service providers to be active in ICT standardisation; in many cases just being an "observer" may suffice to get early warnings in case a new technology emerges as potential competitor.

DongBack's book discuses all the issues I have touched upon thus far and more. I am convinced that the book will be highly beneficial for readers with very different backgrounds. For one, researchers may use the book's tools and insights, for example, to extend the framework for an organizational standards strategy. The book is also relevant for corporate standardisation managers as well as for general strategists. After having read the book, they will be in a much better position to adequately appreciate the importance of ICT standards and standardisation and to devise an adequate corporate standardisation strategy and to actually implement it. This, in turn, may ultimately help a firm to survive.

All in all, I trust the book will significantly contribute to a better understanding of how corporate standardisation strategies are shaped and thus, ultimately, to better strategies. At the end of the day, all of us would stand to benefit from that.

Kai Jakobs
Aachen University, Germany

Kai Jakobs *joined RWTH Aachen University's Computer Science Department as a member of technical staff in 1985. Since 1987, he has been Head of Technical Staff at the Chair of Informatik 4 (Communication & Distributed Systems). He holds a PhD in Computer Science from the University of Edinburgh and is a Certified Standards Professional. Kai's current research interests and activities focus on various aspects of ICT standards and the underlying standardisation process. He is Vice President of the European Academy for Standardisation (EURAS), as well as founder and editor-in-chief of the International Journal on IT Standards & Standardization Research, and of the Advances in Information Technology Standards and Standardization Research, and the EURAS Contributions to Standardisation Research book series. He has (co)authored/edited a text book on data communication networks and, more recently, sixteen books on standards and standardisation processes, with a focus on the ICT sector. More than 180 of his papers have been published in conference proceedings, books, and journals. He has been on the programme committee and editorial board of numerous international conferences and journals and has served as an external expert on evaluation panels of various European R&D programmes on both technical and socio-economic issues.*

Preface

OVERVIEW OF THIS BOOK

The term "standard" is commonly used in daily life with various meanings, such as reference, minimum quality, and the specification of compatibility (interface) between components. These variations in meaning likewise apply for the term "technology standard." Research on technology standards first grew in the 1980s, examining the roles and types of standards used in the rapidly growing Information and Communications Technology (ICT) sector (David & Greenstein, 1990). Due to the intense need for compatibility and component integration necessary for a product or service to function and compete in the ICT sector, technology standards play a highly significant role. Broadly speaking, many factors enter into the adoption of technology standards including market competition, government regulations, and alliances and collaborations among organizations (Bores, et al., 2003; Chiesa & Toletti, 2003; Yoffie, 1996). Thus, one of the key goals of this book is to develop a framework for sorting out the factors that contribute to the establishment of compatibility (interface) standards in the ICT sector.

Standards strategy is an organization's long-term plan to achieve its goals by using standards to gain or sustain competitive advantage. An organization that first develops a standard or adopts it early on can gain great competitive advantage. It can also sustain competitive advantage in the post-standardization period, after a standard has been widely established, by locking-in users because of greater economy of scale. To properly attain such competitive advantages, an organization must create specific strategies for developing, promulgating, and implementing standards (Grindley, 1995; Tassey, 2000). The subject of this book is the process of such deliberate standards strategy setting within the context of an organization's overall competitive strategies.

There have been three main strands in the body of literature on technology standards. One focuses on the economic factors involved in spreading a technology to make it into a standard—economies of scale, network externality, the bandwagon effect, etc. (e.g. Chen & Forman, 2006; David & Steinmueller, 1994; Katz & Shapiro, 1985; Liebowitz & Margolis, 1994). The second strand considers the agents of standardization—whether standards emerge without an identified originator, from government regulation, or by deliberate action of organizations working singly or together (e.g. Besen & Farrell, 1994; Chiesa & Toletti, 2003; David & Greenstein, 1990). The third trend in the literature focuses on the role of IPR (Intellectual Property Rights) in standards strategies—for example, how organizations weigh protection of their rights with the necessity of opening up their technology in order to spread it (e.g. Blind & Thumm, 2004; Lea & Hall, 2004; Lemley, 2002).

However, what is not clear from the literature is how the three strands of aspects are to be integrated and synthesized so as to formulate concrete standards strategies. In other words, the existing literature fails to give an account of why organizations choose certain strategies. Moreover, many interesting questions have not been addressed. For example, how does any given actor decide how to stimulate the necessary economic factors to spread its technology, how to create an effective network using that technology, whether to work with collaborators or go it alone, and whether to close or open IPR? What influences these decisions? What elements in an actor's situation will lead it to make one choice or another?

This book proposes to go beyond a mere listing or categorization of the types of standards strategies and try to understand why organizations choose certain strategies. Therefore, the first part of the research question is: *What is the nature of organizational strategies for technology standards?* To answer this question, it is necessary to review and draw upon how previous researchers frame their analyses of standards strategies. Then a framework that is rich enough to answer the main question of this research is formulated: *How do organizations reach and then adapt their strategies for standards before, during, and after the industry-wide standardization process?*

To be able to address this broad research question, a holistic perspective that is comprehensive enough is needed to analyze the *context* of organizational standards strategy, including the situations of the organizations involved and how these organizations interpret their situations and develop their particular strategies. While previous research draws on established theories such as Game Theory and economic theories of Industrial Organization, the framework of this research is based on the combination of two newer theories—Actor Network Theory (ANT) and the principles of Self-Organized Complexity.

According to Kaghan and Bowker (2001), an actor network is defined as "any collection of human, non-human, and hybrid (human/non-human) actors who jointly participate in some organized (and identifiable) collective activity in some fashion for some period of time" (p. 258). The network is not always intrinsically coherent and can include conflicts among actors. For this reason, the network can change and fall apart over time. ANT is useful because it takes into account the complexity of real life and can explain the interactions of organizations. However, ANT does not account for the context in which actors make decisions. Nor does it account for the unfolding of the network—why do certain networks emerge, and how do they emerge? (Gao, 2005). In order to answer these questions, it is necessary to analyze the larger picture of industry-wide dynamics and the patterns of how many organizations interact. For this purpose, the theory of Self-Organized Complexity is drawn.

Self-Organized Complexity, which is also called Complexity Theory or Self-Organization Theory, focuses on self-organization in complex phenomena. Self-organization occurs when a configuration or pattern emerges from the interaction of various independent actors over time, without the intervention of a central controller (Drazin & Sandelands, 1992; Anderson, 1999). ICT standardization exhibits the basic characteristics of self-organized complexity, namely the existence of numerous actors and the emergence of configurations or patterns as a result of the collective behavior of interacting actors (Egyedi, et al., 2007). Through the interactions of the strategies and tactics of the organizations striving for technology standardization, certain patterns of technology standards emerge in an organized manner—perhaps developed through voluntary standard-setting collaboration by various firms, perhaps as a standard that emerges through market competition.

This book proposes to integrate the two theories, Self-Organized Complexity and ANT, to create a theoretical approach that seems promising for providing a holistic analysis of the process of technology standardization. Self-Organized Complexity provides us with a way to analyze the overall phenomenon of technology standardization (Anderson, 1999), because it accounts for the existence of many actors, their isolated or interdependent actions, and the myriad interactions that form recognizable configurations and patterns. Complementing this, ANT can help us understand the dynamics of the actors' interactions that ultimately lead to the emergence of technology standards. Thus, the combination of these two theories should allow us to explain not only the formulation of standards strategies of individual organizations, but also how their strategies interact, and consequently, the emergence of standards and their unexpected side-effects.

Specifically, the following general "picture" of the standardization process is proposed:

1. The existing configurations, trends, and patterns in technologies and markets prescribe the situation of each individual organization.
2. Each organization interprets its situation through its orientation, values, and goals.
3. Each organization creates its standards strategies based on this interpretation of its situation and takes actions to pursue its strategy.
4. These actions of individual organizations interact to create or shape a network of relations and value creation.
5. These networks of relations and value creations "self-organize" into patterns of technology standards adoption and ultimately give rise to new structures or configuration of the industry.
6. The newly emerged configuration displaces the old, and the whole process (1) through (5) repeats, thereby continuously unfolding and reshaping industry structure as generations of technologies and markets.

In order to facilitate more systematic analyses of (a) organizational situations, (b) organizational interpretation of situations, and (c) organizational choices of standards strategy, specific key elements and factors that characterize these three aspects of the organizational standards strategy process are identified. These elements and factors are derived from a review of the literature, and are consistent with the perspective of value creation and capture, which is the foundation for theories of organizational strategy (Lepak, et al., 2007).

With respect to the pursuit of technology standards, at least six fundamental factors of an organization's situation may be identified:

1. An organization's capabilities to meet market needs and opportunities.
2. The availability of complementary products or compatibility of products/services in the market.
3. The innovativeness of the technology involved.
4. The relative position of the organization in the market.
5. The availability of alternative or substitutable technologies in the market.
6. The characteristics of intellectual property rights regarding the technology involved.

The key elements that constitute organizations' interpretations of their situations are much more difficult to characterize. Logically, it is impossible for an outside investigator to fully describe and understand an organization's subjective interpretation of its situation. The outside investigator must, so to speak, interpret the actors' interpretations. It is implicitly hypothesized that the only possible method of reasonable and workable interpretation is from the perspective of value. Specifically, it is hypothesized that an organization always evaluates its situation from the perspective of self-interest, and from the angle of creating and capturing value; for example, what value its resources and capabilities might add or create to existing products and services; whether its value-adding activities can be sustained in potential future markets; whether its situation is advantageous for creating or capturing value for emergent market needs and opportunities; etc. All in all, an organization's interpretation of its situation is to be imputed as its particular self-interested responses to the above six factors that characterize its situation.

Lastly, this book suggests to analyze organizations' "space of maneuver" in choosing their standards strategies and the interaction of their strategies with regard to three different but complementary aspects: (1) the possible configuration of the emerging value network, (2) the shaping of venues where technology standards may arise (e.g. consortium or standard setting organizations), and (3) the competitive advantages gained or lost as the result of the openness of IPR (Intellectual Property Rights) necessary for the adoption of a technology standard.

The suggested framework will first be tested with relatively two simple "prototype" cases—the standardization of PC architecture in the USA and in Japan—to illustrate how the theoretical framework is to be applied (see Appendix). The main case study in this book considers the evolution of mobile communications technology standards, from the first generation (1G) through the second generation (2G) to the third generation (3G). The mobile communications industry is becoming a bigger and bigger sector of the information and communications industry. According to the 2006 OECD Information Technology Outlook, companies that produce mobile communications equipment and services create approximately 40% of revenues in the whole ICT industry. The development of the mobile communications industry involves many competing organizations (i.e. actors in ANT). These different organizations, according to their IPR ownerships, industry positions, political affiliations, etc. propose different mobile technology standards that would enhance their perceived profit opportunities. The goal of the case study is to describe this very complex inter-organizational co-opetition vying for control of mobile communications technology standards. Going beyond mere description, the case study analyzes how the organizations' standards strategies interact and how these interactions "self-organize" the whole mobile communications sector into certain discernible configurations. This case study also traces and analyzes how the emerged technology and business configurations unfold over time from 1G through 2G to 3G as the result of decades-long standards strategies of the influential organizations.

Because of its unavoidably broad scope, this case study considers the proposal, modification and adoption, and implementation of many mobile communications technology standards. The study of each of these standards, such as NMT, AMPS, GSM, IS-95 (cdmaOne), WCDMA, and CDMA2000, may be considered as a stand-alone sub-case. The challenge for this case study is to integrate these sub-cases to give a comprehensive understanding of the unfolding of the mobile communications industry in the past three decades.

This book is organized as follows: First, an overview of the received definitions and classifications of standards is presented, followed by a review of the specific aspects of the ICT (Information and Communications Technology) industry peculiar to the establishment of standards. After a brief discussion of standards strategy in the general business environment, the special role of standards strategy in the ICT industry is surveyed by reviewing the existing literature.

Next, the theoretical background of this research is developed, first by reviewing existing theories that researchers have used in their research on standards, then by critically examining the research questions they address. It is then argued that existing research does not truly analyze how concrete and specific organizational strategies interactively shape the emergence of standards, and in order to do a proper analysis of the phenomenon of standards in the ICT industries, an integrated theoretical framework (based on the newer theories of ANT and Self-Organized Complexity) is necessary. The theoretical framework for this book is then formally developed. Using this framework as context, the research questions are formally stated, and the specific method and approach to answer the research questions is proposed and justified.

Finally, the proposed theory method and approach is applied to analyze the decades-long development of the mobile communications industry as driven by the unfolding of mobile technology standards.

In summary, this book consists of two parts: (1) the development of a theoretical framework for the description and analysis of technology standardization and organizational standards strategy, and (2) the application of the proposed framework to guide a substantial case study, with the purpose of validating the appropriateness of the proposed framework for the study of standardization in general.

The potential contributions of this book are these: (1) developing a holistic theoretical framework for understanding the complex phenomenon of standardization by integrating two theories (Self-Organized Complexity and Actor Network Theory); (2) understanding how organizations determine their standards strategies; (3) analytically comprehending the evolution of mobile communications technology standards; (4) recognizing the importance of organizational standards strategy as a way for organizations to gain or sustain competitive advantage; and (5) highlighting the field of organizational standards strategy as an important part of overall business strategy.

Different from other mobile related books that usually focus on a specific technology, this book can be read by many readers from various areas. The target audience includes CIOs, managers, technology strategists, policy makers, engineers, and experts in universities, business schools, companies, and governments, who deal with the topics:

- ICT Standardization
- Standards Strategies
- New Product/Service Development
- Technology Transfer, Marketing, and Commercialization
- Technology Foresight and Forecasting
- Information and Communications Technology Management
- The Integration of Technology and Business Strategies
- R&D Management
- Technological Alliances, Mergers, and Acquisitions
- Systems Dynamics Modeling
- Social Networks

The readership would consist of graduate students or professors in management of technology or MBA programs as well as scientists, engineers, and various experts working in the mobile communications industry. The book can be used in these fields as supplemental reading material and as an important reference to many researchers.

DongBack Seo
Hansung University, Repbulic of Korea

REFERENCES

Anderson, P. (1999). Complexity theory and organization science. *Organization Science, 10*(3), 216–232. doi:10.1287/orsc.10.3.216.

Besen, S. M., & Farrell, J. (1994). Choosing how to compete: Strategies and tactics in standardization. *The Journal of Economic Perspectives, 8*(2), 117–131. doi:10.1257/jep.8.2.117.

Blind, K., & Thumm, N. (2004). Interrelation between patenting and standardisation strategies: Empirical evidence and policy implications. *Research Policy, 33*(10), 1583–1598. doi:10.1016/j.respol.2004.08.007.

Bores, C., Saurina, C., & Torres, R. (2003). Technological convergence: A strategic perspective. *Technovation, 23*(1), 1–13. doi:10.1016/S0166-4972(01)00094-3.

Chen, P. Y., & Forman, C. (2006). Can vendors influence switching costs and compatibility in an environment with open standards? *Management Information Systems Quarterly, 30*, 541–562.

Chiesa, V., & Toletti, G. (2003). Standard-setting strategies in the multimedia sector. *International Journal of Innovation Management, 7*(3), 281–308. doi:10.1142/S1363919603000829.

David, P. A., & Greenstein, S. (1990). The economics of compatibility standards: An introduction to recent research. *Economics of Innovation and New Technology, 1*, 3–41. doi:10.1080/10438599000000002.

David, P. A., & Steinmueller, W. E. (1994). Economics of compatibility standards and competition in telecommunication networks. *Information Economics and Policy, 6*(3-4), 217–241. doi:10.1016/0167-6245(94)90003-5.

Drazin, R., & Sandelands, L. (1992). Autogenesis: A perspective on the process of organizing. *Organization Science, 3*(2), 230–249. doi:10.1287/orsc.3.2.230.

Egyedi, T. M., Vranchen, J. L. M., & Ubacht, J. (2007). *Inverse infrastructures: Coordination in self-organizing systems.* Paper presented at the 5th Conference on Standardization and Innovation in Information Technologies (SIIT). Calgary, Canada.

Gao, P. (2005). Using actor-network theory to analyse strategy formulation. *Information Systems Journal, 15*(3), 255–275. doi:10.1111/j.1365-2575.2005.00197.x.

Grindley, P. (1995). *Standards, strategy, and policy: Cases and stories.* Oxford, UK: Oxford University Press. doi:10.1093/acprof:oso/9780198288077.001.0001.

Kaghan, W. N., & Bowker, G. C. (2001). Out of machine age? Complexity, sociotechnical systems and actor network theory. *Journal of Engineering and Technology Management, 18*(3/4), 253–269. doi:10.1016/S0923-4748(01)00037-6.

Katz, M. L., & Shapiro, C. (1985). Network externalities, competition, and compatibility. *The American Economic Review, 75*(3), 424–440.

Lea, G., & Hall, P. (2004). Standards and intellectual property rights: An economic and legal perspective. *Information Economics and Policy, 16*(1), 67–89. doi:10.1016/j.infoecopol.2003.09.005.

Lemley, M. A. (2002). Intellectual property rights and standard-setting organizations. *California Law Review*, *90*(6), 1889–1980. doi:10.2307/3481437.

Lepak, D. P., Smith, K. G., & Taylor, M. S. (2007). Value creation and value capture: A multilevel perspective. *Academy of Management Review*, *32*(1), 180–194. doi:10.5465/AMR.2007.23464011.

Liebowitz, S. J., & Margolis, S. E. (1994). Network externality: An uncommon tragedy. *The Journal of Economic Perspectives*, *8*(2), 133–150. doi:10.1257/jep.8.2.133.

OECD. (2006). *Information technology outlook*. Paris: OECD.

Tassey, G. (2000). Standardization in technology-based markets. *Research Policy*, *29*(4-5), 587–602. doi:10.1016/S0048-7333(99)00091-8.

Yoffie, D. B. (1996). Competing in the age of digital convergence. *California Management Review*, *38*(4), 31–53. doi:10.2307/41165853.

Acknowledgment

This research was financially supported by Hansung University. The author would like to thank Dr. Mak, whose comments helped improve the book significantly.

Chapter 1
Background of
Standards Strategy

EXECUTIVE SUMMARY

First, the literature of research on standards is reviewed, and an overview of the definitions and classifications of standards is provided. Then particular aspects of the ICT (Information and Communications Technology) industry as related to standardization are reviewed. Finally, standards strategies in the ICT industry are examined by critically reviewing the existing literature and identifying important areas that need further investigation.

1. DEFINITION OF STANDARD

In this section, various received definitions and categories of standards will be presented, in order to arrive at a definition of standard, which will be suitable for the purpose of this study.

The term "standard" is commonly used in daily life with various meanings. While it can be used to refer to various concrete things—including the banners of the Roman legion, a structure built for or serving as a support, and tree with an erect stem—it is most commonly used to designate "a means of determining what a thing should be," or more specifically:

DOI: 10.4018/978-1-4666-4074-0.ch001

- Something established by authority, custom, or general consent as a model or example.
- Something set up and established by authority as a rule for the measure of quantity, weight, extent, value, or quality (Merriam-Webster's Dictionary, p. 1148).

Research on standards first appeared in the 1980s, in the examination of the roles and types of standards used in the information sector (David and Greenstein, 1990). David and Greenstein (1990) gave a survey of this research. Of the studies they examined, few actually provided a concrete definition for the term "standard." In their stead, David and Greenstein proposed a general definition for technical standard: "a set of technical specifications adhered to by a producer, either tacitly or as a result of a formal agreement" (p. 4).

This definition has been cited by many researchers since then (e.g. Chiesa and Toletti, 2003). Another commonly used definition is the one proposed by Tassey (2000). He suggests that "a standard can be defined generally as a construct that results from reasoned, collective choice and enables agreement on solutions of recurrent problems" (p. 588), and he further specifies industry standards as "a set of specifications to which all elements of products, processes, formats or procedures under its jurisdiction must conform" (p. 588) (e.g. van Wegberg, 2004).

Different types of standards may be categorized by their characteristics (David and Greenstein, 1990; Grindley, 1995; Tassey, 2000). The literature on standards offers different ways to classify them according to their purpose or the role they play (David and Greenstein, 1990; Tassey, 2000). David and Greenstein (1990) distinguish between standards as model reference, as minimal level of quality, and as criteria of compatibility (interface) between components. Reference standards and minimal quality standards represent specifications that a product as a whole should conform to, while compatibility (interface) standards focus on specifications for the relationships between components in order for them to function as one system.

Grindley (1995) discerns a similar split and classifies standards into two types: those that control the quality of a product as a whole entity, and those required for compatibility of products as parts in larger systems. Quality standards, focusing on the features of the product itself, are subdivided into two types: minimum attributes related to the measurement, gradation, and public regulation of product performance as whole entities; and product characteristics as related to style, tastes and production economies. Compatibility standards, concerning the relationship among whole products or services, address three areas: complementary products, complementary services, and direct networks. Complementary products refer to items that perform complementary functions, for example, electronic component systems for automobiles, and software applications for computers. They are frequently supplied by separate producers or even separate industries. Complementary services usually refer to supporting services; for example, maintenance service for automobiles or computers, management service for computer facilities, and training service for users. Direct networks refer to the ways and means for connecting users to the same core product or service; for example, railway routes and telecommunications links.

Tassey (2000) offers a more detailed taxonomy along the same vein, classifying standards based on the functions for which they are generally used. First, there are standards that play a role in specifying quality and reliability, and in measuring product or service performance. For example, when firms purchase parts for their products, they require a specific range of characteristics such as elasticity and durability of parts. Second, there are standards that inform the specification and evaluation of technical and engineering designs, for example the standardized measurements used in semiconductor fabrication facilities. Third, there are standards for ensuring the compatibility of the function and performance of complementary systems, and standards for providing technology specifications to ensure the interoperability of component systems. Fourth, there are standards whose adoption serves to reduce variations in the characteristics of a product, for example in size and quality, in order to reach economies of scale.

Among the many types of standards, those used for the purpose of compatibility and interoperability have received the most attention from practitioners and researchers in the field of information and communications technology (David and Greenstein, 1990; Grindley, 1995; Tassey, 2000; van Wegberg, 2004). It is a characteristic of the ICT industry that products and services always involve many complementary subsystems, components, and complementary services (David and

Greenstein, 1990; Grindley, 1995; Tassey, 2000; van Wegberg, 2004). Therefore, compatibility standards that specify how myriad parts should function together are of fundamental importance to the ICT industry, and it is of strategic importance for organizations in the industry to attempt to shape and control compatibility standards. Because of their importance, this study will focus on standards for compatibility and interoperability, while not ignoring other types of standards. Drawing on the reviewed literature, and focusing on compatibility, the definition of standard used in this book is taken to be:

- **Standard:** An agreed upon technological format, specification, or architecture to ensure compatibility within or among components, systems, products or services so as to make them work together complementarily.

2. INFORMATION AND COMMUNICATIONS TECHNOLOGY

In this book, we are concerned with standards in the Information and Communications Technology (ICT) sector. To understand the roles of standards in the ICT industry, the nature of the ICT industry is first examined. The definition of ICT is adopted, which was proposed and agreed upon by the members of Organization for Economic Co-operation and Development (OECD) in 1998, which was based on the International Standard Classification of business activities (ISIC Rev. 3). OECD, founded in 1961, is an international organization that supports the building of strong economies in its member countries. According to this agreement, the ICT sector is defined as "a combination of manufacturing and services industries that capture, transmit and display data and information electronically" (OECD, 2002, pp. 81). It also specifies the functionalities that products and services should carry in order to be included in this definition:

Products "*must be intended to fulfill the function of information processing and communication including transmission and display,*" or "*must use electronic processing to detect, measure and/or record physical phenomena or control a physical process.*" Services "*must be intended to enable the function of information processing and communication by electronic means*" (OECD, 2002, p. 81).

To appreciate the scope of the ICT sector thus defined, listed below are the business categories included in the ICT sector, as based on the ISIC Rev. 3 (OECD, 2002, p. 81):

- Manufacturing
  - **3000:** Office, accounting and computing machinery.
  - **3130:** Insulated wire and cable.
  - **3210:** Electronic valves and tubes and other electronic components.
  - **3220:** Television and radio transmitters and apparatus for line telephony and line telegraphy.
  - **3230:** Television and radio receivers, sound or video recording or reproducing apparatus and associated goods.
  - **3312:** Instruments and appliances for measuring, checking, testing, navigating and other purposes, except industrial process equipment.
  - **3313:** Industrial process equipment.
- Services
  - **72:** Computer and related activities.
  - **5151:** Wholesale of computers, computer peripheral equipment and software.
  - **5152:** Wholesale of electronic and telecommunication parts and equipment.
  - **7123:** Renting of office machinery and equipment (including computers).
  - **6420:** Telecommunications.

It can be seen that the ICT sector covers a wide area: from physical media, including networks and devices to transmit audio-visual and text data, to retail and services to deliver those data. Basically, every manufacturing or service industry related to information processing and data delivery, regardless of data type, is a part of the ICT sector.

The reason the ICT sector includes so many industries is that a complete product or service in the ICT sector usually requires a variety of complementary products and/or services (Bores et al., 2003; Gambardella and Torrisi, 1998). For example, computer hardware without software is useless. Likewise, a mobile communication handset becomes valuable only when it is connected to network service provided by a service provider; a television set is only useful when it can receive data from TV broadcasters. The fact that ICT products and services are built up from component systems and services implies a great need for compatibility between different types of networks, between personal devices and network equipment, and between software and hardware, so as to provide comprehensive seamless goods and services (Bores et al., 2003; Gambardella and Torrisi, 1998).

Due to the complex technical complementarity required in the industry, players in the ICT sector have to somehow agree on certain technological formats, specifications, and architectures in order to achieve compatibility of products and services, and interoperability of component systems (Bores et al., 2003; Yoffie, 1996). The outcome of any agreed-upon technological format, specification, or architecture is a standard, whether it comes out of market competition, or out of some co-opetitive struggle among business organizations before commercialization (Chiesa and Toletti, 2003).

The fundamental role of technology standards in the ICT industry is the first reason for focusing on the ICT sector for this research in standards strategy as organizational strategy. The second reason is that the ICT sector has become the driver of socio-economic changes and the basis

of productivity growth (OECD – Science and Technology, 2006; Powell and Snellman, 2004; Tripathi, 2006). Indeed, the current development of the ICT industry is driving the building of new infrastructure for all other industries (Powell and Snellman, 2004; Tripathi, 2006). Standardization of technology and globalization in business seem to go hand in hand. Therefore, it makes sense to start with the ICT sector if one intends to investigate the role and impact of standards strategy on organizational strategy in general.

3. STANDARDS STRATEGY

Because standards can play an important role in business success, organizations may use standards as a strategy to gain a more advantageous position. To do this deliberately, they may create strategies for developing, promoting, and diffusing particular standards that enhance their competitiveness (Grindley, 1995; Tassey, 2000). In this chapter, the meaning of standards strategy in general business environments will be introduced before discussing the specific aspects of standards strategy in the ICT sector.

Elements of strategy are in fact implicitly embedded in the very definition of standards. Through reviewing the literature, three commonalities in the various definitions of standards have been identified:

1. A problem arises due to the lack of a common process, product, or unit of measurement among some group of actors.
2. Alternative (competing) options for a solution to the problem are proposed.
3. The actors go through certain processes of cooperative negotiation and/or competitive strife until certain solutions are agreed upon.

These elements can be discerned in the definition of standards offered by David and Greenstein (1990), "a set of technical specifications adhered to by a producer, either tacitly or as a result of a formal agreement" (p. 4); and by Tassey (2000), "a standard can be defined generally as a construct that results from reasoned, collective choice and enables agreement on solutions of recurrent problems" (p. 588). Tassey (2000) further specifies industry standards as "a set of specifications to which all elements of products, processes, formats or procedures under its jurisdiction must conform" (p. 588). The standard can be perceived as an agreed-upon solution that provides the lacking common specifications, solving the problem at hand. Often, the problem is easily identified, and viable solutions may be readily formulated. However, for a particular common solution to be adopted, there is the problem of reaching agreement among the involved actors. Coming to agreement could be a very difficult task, if the parties involved have their own technology or process, and if it is in their interest to insist on their own way of doing business. Below is a simple illustrative example.

It is problematic when individuals, various groups of people, or organizations attempt to communicate with each other without a common process and the compatible products, performance measurement, and interfacing rules to support the shared process. Let us imagine that region A uses postal codes of six digits, where the first three digits designate an area and the last three digits designate specific blocks within the area. Let us say that in region B, the reverse is true – the first three digits in region B's postal code designate specific blocks and the last three digits designate an area. This is not a problem as long as there is no mail exchange between the two regions. However, it is a problem if people in the two regions desire to correspond with each other, especially if users do not acknowledge the difference in advance.

Once the two regions discover there is a difference in their postal codes, they need to solve this problem of incompatibility. There are several ways to solve the problem: (1) one region adopts the protocol of the other region; (2) both regions develop and adopt a new common postal code protocol together; or (3) they can develop and implement an interface which provides an agreed method to convert the postal codes when mail moves between regions. The interface could be implemented in at least three different ways. One method is to educate the users to write the postal code in the format of the recipients' region. The second method is to train carriers to translate postal codes of mail moving between the two regions. The third way is developing a machine that can automatically convert postal codes depending on the destination and print it on the envelope.

The actors involved, in this case the two regions, need to agree on either standardizing the postal code or adopting a workable interface. Having one region adopt another region's protocol might not appear too difficult. However, the two regions may be rivals or their regional pride might be too strong to adopt another's postal code.

This is a very simplified example involving only two regions. In a more realistic situation, there are many regions with incompatible postal codes, which nevertheless desire to exchange mail. In this case, it might be more economical in the long run to adopt a universal standardized postal code instead of building complex interface systems to accommodate the many streams of mail exchanges. But, in the short term, it might simply be too expensive to switch all one's existing equipment and facilities to accommodate a new universal standard. Pride of ownership also factors into the attitude of a region towards the ways and means of adopting a standard. In general, each actor naturally prefers to standardize to a process, product, and protocol that would benefit it more and cost it less, and each will therefore pursue a standards strategy according to how it interprets the situation and perceives its interest. The outcome cannot be pre-determined without due consideration of the concrete process of standardization put into play.

Most definitions of standards in the literature focus on the problem and the solution aspects (e.g. David and Greenstein, 1990; Tassey, 2000), but the third aspect – the process of agreeing on the solution – is perhaps more critical, though ignored. In the real world, the actors (or organizations) must somehow agree on a solution while adjudicating and trading off a whole host of complementing and conflicting interests as driven by various business costs, values, and opportunities (Grindley, 1995; Tassey, 2000).

One example of how conflicting interests arise in the process of setting standards is the standardization of railway track gauge in North America. In the 1800s, different railway companies in North America built railways with different track gauges (Hilton, 1990; Puffert, 2000; Puffert, 2002). The incompatibility of gauges presented no problem when the railway companies operated in disjoint regions of the continent, but it became a huge obstacle when the companies expanded their tracks to other regions. Trains could only operate on the tracks with the gauge for which they were built. This fact precluded interoperability of trains and rail shipments. The obvious solution to this problem of lack of interoperability was to standardize rail gauges. However, each railway company was strongly interested in making its track gauge the standard – any company that would lose this battle would have to replace all its tracks as well as the moving stocks, thus incurring very high switching costs. Eventually, the 4-feet 8.5-inch gauge became the standard of North American railways. This came about mainly because this particular gauge had been extensively adopted earlier by rail companies in substantial regions of the country such as the Northeast, mid-Atlantic, and Midwest (Puffert, 2000; Puffert, 2002), and because of the strategic formation of a coalition of these companies.

As the above railway example illustrated, which standards are adopted and how they are adopted can have a huge impact on a company's growth and prosperity. Therefore, organizations need to take standards into account in their organizational strategies (Grindley, 1995; Bekkers, 2001). In accordance with the definitions of organizational strategy found in the literature – for example, "determination of the basic long-term goals of an enterprise, and the adoption of courses of action and the allocation of resources necessary for carrying out these goals" (Chandler, 1962, p. 13) and "either the plans made, or the actions taken, in an effort to help an organization fulfill its intended purposes" (Miller and Dess, 1996, p. 38) – standards strategy may be defined as "an organization's long-term plan to achieve its goals or intended purposes concerning the establishment and/or adoption of standards." Below, how standards strategy fits within the overall context of organizational strategies will be discussed.

For a corporation that has multiple business units, there are three levels of organizational strategy (Kay et al., 2003). A good example of a company with multiple business units is GE; it has six major business units – GE Capital, GE Lighting, GE Healthcare and Medical, etc. The first level of organizational strategy is the level of corporate-level strategy, which concerns "how to identify the businesses that should form a core portfolio for a corporation and how to find ways of adding value to those businesses" (Goold and Luchs, 2003, p. 29). The second level is that of business or competitive strategy, which concerns how the business units deal with issues such as "how to configure value chain optimally, what products and services to offer to what specific market segments, how to achieve differentiation from the offerings of competitors, and how to control costs in order to be able to be price-competitive" (Faulkner and Campbell, 2003, p. 12). The third level of organizational strategy consists of various sub-strategies that concern functional issues such as manufacturing, operation, finance, and marketing (McLaughlin et al., 1991). These sub-strategies serve both as constraints and supports to the business units in the implementation of their competitive strategy into activities. There can be many inter-related sub-strategies coordinated in such a way as to support the overall business or competitive strategy (McLaughlin et al., 1991). For example, Fifield (1992) defines marketing strategy as "the process by which the organization translates its business objective and business strategy into market activity" (p. 13) and introduces a variety of strategies in this category such as target market strategies, target market needs, and market positioning. He emphasizes that all of these strategies included in marketing strategy should support the overall organizational business or competitive strategy in interaction with sub-strategies from other divisions.

Given this three-level structure of organizational strategies, the question is where does standards strategy belong? Is standards strategy a separate sub-strategy based on its functionality, just like marketing or operations strategies? Is it a part of one of the functional strategies? Or is it in itself a competitive strategy requiring support from other sub-strategies?

To help ascertain the status of standards strategy, let us consider again the example of standardization of railway gauge in North America. When rail companies grew geographically and started expanding into overlapping territories, it was more effective and economical to use other companies' existing tracks instead of laying down their own tracks (Puffert, 2000, 2002). Thus, the standardization of railway gauges would enable operational efficiency and business effectiveness, for the individual railway company and for the overall industry. However, from the perspective of individual train companies, this standardization battle over railway gauge was not so straightforward. The failure to make one's own gauge the standard would create incredible switching costs and require extensive changes in all divisions – operations, finance, track manufacturing and installation, with no exception (Puffert, 2000, 2002). Thus, the organizational standards strategy in this case was not simply a part of operations

strategy, but rather something that necessarily encompassed several divisions of sub-strategies.

One example is Wal-Mart's efforts to standardize its Radio Frequency Identification (RFID) technology as an industry standard. Although RFID technology provides efficiency and effectiveness in operations and supply chain management, its development and implementation goes beyond any single division of Wal-Mart and requires the involvement of finance, logistics and supply chain, operations, and information systems. The process of developing RFID, from conception to standardization, requires not only significant financial resources but also personnel commitments and cooperation from Wal-Mart's different divisions. That is why Wal-Mart Stores' RFID Strategy Team was created across divisions specifically to develop, install and standardize its RFID technology (Ferguson, 2006; Sullivan, 2004).

As brought out in these examples, the impact and process of standards strategy straddle two or more sub-strategies of functional divisions. Moreover, standards strategy should also address the problem of configuring the value chain more effectively and efficiently to gain and sustain competitive advantage, which, according to Faulkner and Campbell (2003), is one of the components of business or competitive strategy. It is therefore argued that standards strategy should be regarded as a separate business or competitive strategy within the overall corporate strategy if the organization has more than one business unit.

The standards strategies of organizations as a part of their competitive strategies interact in the market. The outcome in general is the adoption of certain standards and the declaration of winners and losers. In other word, the market somehow collectively settles into certain configurations of technology and business relations. To the extent that organizational standards strategies should anticipate the ultimate emergence of industrial configuration, standards strategies should be considered as a long-term competitive strategy.

Having placed standards strategy in the context of overall organizational business or competitive strategy, it is now turned to the question of what the substance of standards strategy is, and how it can help organizations pursue competitive advantage.

Standards strategy is an organization's long-term plan to achieve its goals or intended purposes by using standards to provide competitive advantage. An organization that first develops a standard or adopts it early on can gain great competitive advantage. For example, the organization can gain early market penetration and market share as a first mover; it can impose switching costs on its competitors as in the case of railway gauge; and it can create lock-in effects for customers, which furthermore influences customer adoption of future products through path dependence (Bekkers, 2001; Grindley, 1995; West and Dedrick, 2000).

Organizations can also sustain competitive advantage in the post-standardization period, after a standard has been widely established, by locking-in their adopters through greater economy of scale (Bekkers, 2001; Grindley, 1995; West and Dedrick, 2000) as the consequence of the network effect (Katz and Shapiro, 1994). One example of this is Microsoft's Windows operating system, which affects users' choice of software products because of the wide range of options available on Windows. However, if it is not the focal point of the locking-in, an organization which gained competitive advantage in the standardization process can lose its advantage in the post-standardization period. For example, IBM standardized its PC architecture quickly and gained significant market share, but it could not sustain its competitive advantage due to the fact that the PC architecture was open and therefore other manufacturers could easily develop PC clones (Grindley, 1995; West and Dedrick, 2000). Thus, organizations need standards strategy not only during the standardization process but also in the post-standardization period to protect their position in the value network so as to sustain competitive advantage.

Organizations that do not gain or sustain competitive advantage due to non-existent or insufficient standards strategy may be able to sustain themselves with a small but steady market share; for example, Apple's OS. Or they may find that they need to abandon their technology and adopt the standardized technology; for example, Sony's Beta videocassette format. This applies not only to organizations that have developed their own technology but also to adopters of these technologies (Grindley, 1995).

4. STANDARDS STRATEGY IN THE ICT SECTOR

Organizations in the ICT (Information and Communications Technology) industry are interested in standards because of the business profits and opportunities standardization can bring. In this regard, organizations in the ICT industry are no different from organizations in other industries. However, there are characteristics unique to the ICT sector that make organizations in the ICT industry more intensely concerned with the process of standardization.

First, ICT goods are complex and require many components and interfaces among them to perform as a system (Boar, 1984; Tassey, 2000). This makes the existence of technological standards highly important and necessary. This also means that potentially many different actors must agree on the standards involved. The value network for any given ICT product or service, which refers to the end-to-end network of actors that provide components, whether it is a vertical or horizontal supply chain, must be configured to deliver a complete system to end-customers (Christensen and Rosenbloom, 1995; Li and Whalley, 2002; Rudberg and Olhager, 2003).

Second, the technology involved in an ICT product or service may be owned by an individual or company through IPR (Intellectual Property

Rights). This means that the holder of IPR for a technology that is adopted as a standard can enjoy tremendous benefits, while competitors or even collaborators must pay for its use (Blind and Thumm, 2004; Cohen et al., 2002; Rivette and Kline, 2000).

Third, the ICT industry moves faster than most other industries. A new ICT product can quickly displace existing products if it brings higher efficiency, more effectiveness, and shorter cycle time (Bores et al., 2003; Fine, 1998). This means that strategies and decisions about technology standards must often be made quickly, and that the impact of making the wrong choice may cause a product or company to become obsolete overnight.

While the literature on ICT standards addresses the first and second characteristics, there is less research examining the impact of the fast speed of evolution in the ICT industry. Organizations that develop or adopt ICT confront the threat of seeing their chosen technology become obsolete quickly due to new substitutable technologies (Bores et al., 2003; Fine, 1998). One way that organizations can deal with this problem is through technology standardization, which can allow organizations to leverage a possible path of market evolution by locking-in users and creating path dependence. However, it is difficult for an organization to judge which technology should be chosen and when to develop or adopt a particular one for standardization when there are several competing technologies (Bores et al., 2003). Determining the right time to participate in a standardization battle depends heavily on an organization's competitive situation. And the timing necessarily affects its standards strategy (This particular issue of timing will be discussed in more detail in the section of Framework for Analyzing Standards Strategy in Chapter 2.).

With respect to the first two distinguishing characteristics of the ICT sector, the research in existing literature follows three broad strands: 1) literature focusing on the mechanics of standards strategies, describing the factors involved in spreading a technology to make it into a standard; 2) literature focusing on organizations' approaches for standardization, specifically in the decision of whether and how to collaborate with other organizations; and 3) more recently, literature focusing on IPR (Intellectual Property Rights) issues in standards strategies.

4.1. First Strand of Literature: Technological Configuration

Without technology compatibilities, it is impossible to begin stimulating economic factors to support standardization (David and Greenstein, 1990). Researchers agree that certain economic factors must exist in all technology standards strategies; specifically, economies of scale must be achieved at some point for standardization processes to succeed (e.g. Schilling, 2002; Yoffie, 1996). Economy of scale refers to the reduction of costs due to greater usage (Chandler, 1990; McGee and Sammut Bonnici, 2002; Sammut Bonnici and McGee, 2002). To gain economies of scale, a technology needs to reach a certain level of installed base, which refers to the number of installations of the technology and the number of users (Farrell and Saloner, 1986; Schilling, 2002). The installed base is supported by complementary products; for example, the installed base of Windows operating systems is mutually reinforced by software created to run on Windows.

Researchers discuss several other important economic factors involved in technology standards strategies. Compatibility is one factor, which has two types as described by Katz and Shapiro (1994). One is horizontal compatibility, which is compatibility between two systems competing for market share. The other is vertical [meaning temporal] compatibility, which is compatibility between successive generations of a similar system. Horizontal compatibility can instigate externality, because firms are less reluctant to produce complementary products when competing systems are compatible with each other. Economists refer to externality as the impact of one's participation on others without compensation (David and Steinmueller, 1994; Katz and Shapiro 1985; Liebowitz and Margolis, 1994). For example, when one company decides to adopt the Internet protocol for its network, it affects others in terms of the growth of the Internet network, because the bigger the network, the more companies there are to communicate with. Once the externality of standards is recognized by many organizations, the coming of the bandwagon effect is just a matter of time. The bandwagon effect refers to the phenomenon of many organizations deciding to adopt a standard in a rapid and cumulative way (Frohlich and Westbrook, 2002; Lee and Chan, 2003; Tsikriktsis et al., 2004). Vertical compatibility can induce path dependence, which refers to whether the decision to pursue a standard is dependent on earlier decisions (Arthur, 1996; David, 1985, 1997; Liebowitz and Margolis, 1990). If the choice of technologies used in earlier generations directly determines choices in later generations, this is called the lock-in effect (Liebowitz and Margolis, 1990). These are some of the more important economic factors that affect technology standards strategies.

These factors are often mutually reinforcing. Farrell and Saloner (1986) introduce the concept of installed base effects in their technological competition model, which leads to lock-in effects and path dependence. Katz and Shapiro (1985) illustrate how network externalities influence compatible standards-setting. Other studies introduce additional factors, such as organizational learning orientation and timing of market entry (Shilling, 2002).

4.2. Second Strand of Literature: Value Network

There has been much research that examines the creation of value networks required to promote technological standards. The researchers analyze the processes and the kinds of collaborations conducive to establishing industrial standards.

David and Greenstein (1990) survey the work of economists on the compatibility of standards and standards-setting processes in the 1970s and 1980s. They create a typology of four different kinds of standardization processes, as categorized by the actors driving the process: (1) un-sponsored standards – those that become standards without an identified originator or sponsoring agency with a proprietary interest, (2) sponsored standards – those that are sponsored by one or more organizations having a direct or indirect proprietary interest, (3) standards agreements within voluntary standards-writing organizations, and (4) mandated standards by government agencies.

The second and third categories—sponsored standards and voluntary standards-writing organizations—are the most discussed in the existing literature, perhaps because they connect directly with organizational strategy. In contrast, the first and fourth categories are less relevant to understanding organizational competition for standards, even though an organization can lobby the government to mandate its technology as a standard.

Other studies explore further subdivisions and hybrids of sponsored and standards-writing processes (e.g. Chiesa and Toletti, 2003; Shapiro et al., 2001). One major subdivision in the sponsored standard category is that between organizations that choose to promulgate a standard alone, and those that choose to do it with others.

The extent to which organizations collaborate with others can also be further subcategorized. For example, Chiesa and Toletti (2003) point out the difficulties encountered when a firm attempts to standardize its technologies all by itself, and introduce three overlapping types of collaborations that organizations deploy to help promote a standard—(1) developing collaborations, (2) sponsoring collaborations, and (3) standard-setting organizations.

A developing collaboration is an agreement between two or more firms to develop a technology together and then make it into a standard, sharing the costs for development and their intellectual proprietary rights. A sponsoring collaboration occurs when a technology has already been developed and organizations work together to gain an installed base and make it the standard. Standard-setting organization refers to an organization that comprises the majority of organizations in an industry that have voluntarily come together to agree on a standard.

Note that a developing collaboration is one form of sponsored standardization. Also, note that a sponsoring collaboration is a hybrid somewhere between individual sponsorship and industry-wide standard-setting organizations. There is a natural correspondence between the categories of collaborations here and the already mentioned categorization of standards by sponsorship.

Chiesa and Toletti (2003) point out that in some cases, although their participation is not mandatory, some organizations may have an interest in not agreeing on a standard and will deliberately delay the process. For each type of collaboration, the researchers discuss the motivations leading organizations to choose it, and the critical factors leading to success of the establishment of a standard, which are related to economic factors like installed base and organizational market position.

Rather than categorizing the various types of collaborations, Besen and Farrell (1994) focus on the prior decision to pursue a standard alone or with others. They explore three different strategic routes that an organization can choose in order to compete in standardization depending on industry structure. First, when organizations have similarly dominant market and technology positions, they may choose to insist on their own distinct technologies, which are incompatible, thus leading to inter-technology competition. Sometimes, in order to increase their market position, they will form coalitions with other organizations to pursue their standard. For example, two coalitions, namely the Blu-ray Disc Association and the DVD Forum, have been competing for the control of the next generation optical disc format. [Blue-ray won in February 2008.] Second, even

when two organizations have similar market and technology positions, they may recognize some benefit to compatibility and settle on one common standard. This standard might be the technology developed by one of the organizations, dictated by the government, sponsored by a third party or specified by a standard-setting organization as a compatible interface. Besen and Farrell (1994) call this situation "intra-technology competition." Third, when there is one dominant player and many minor players, the dominant player will often instigate inter-technology competition through various techniques—for example, by threatening to enforce intellectual property rights, or modify technologies and change interface frequently—thus forcing the minor players to try to make their products at least partially compatible to the standards of the dominant player. The authors draw on Game Theory to explain how organizations logically weigh the payoffs between collaboration and sole sponsorship of a standard in each of the cases described above.

Other researchers follow Besen and Farrell (1994)'s approach in their studies of organizational standards strategies in a competitive market. Oshri and Weeber (2006) investigate the competition and cooperation between organizations in the process of standard-setting. Van Wegberg (2004) presents a contingency framework that identifies the conditions affecting an organization's choice of standardization process. In addition to Besen and Farrell's inter-technology and intra-technology competition, van Wegberg proposes the concept of hybrid standardization processes where organizations may compete in the early life cycle of a product and later cooperate in standard-setting—for example, by sponsoring competing technologies as they jointly develop a compromise. According to van Wegberg, hybrid processes combine the advantages of both market selection and negotiated decision-making. He provides a framework for identifying the mix of conditions, which would lead to the decision to go it alone, to collaborate, or to do some hybrid of both. These conditions include network externalities, intensity of competition, and switching costs. He argues that the main consideration is the degree of modularity in the system technology involved, because greater modularity of information and communication technologies tends to lead to fewer switching costs, which substantially affects organizations' preference for a particular standardization process.

4.3. Third Strand of Literature: IPR Strategy

Intellectual Property Rights (IPR) include patents, copyrights, design rights, trademarks, and trade secrets. The main effect of IPR on the economy is generally thought to be in providing incentives for innovation, thus enhancing economic growth and increasing investment. From an organization's perspective, IPR provides legal monopoly power over a product or a technology, protects distribution of the product, increases the organization's value, and creates a better bargaining position (Blind and Thumm, 2004; Cohen et al., 2002; Rivette and Kline, 2000).

Since the development of innovative technology requires organizations to expend substantial amount of resources, IPR can compensate by providing legal recognition of sole ownership of that innovation for a certain period of time (Blind and Thumm, 2004). Without this recognition, organizations would be discouraged from investing their resources in developing innovative technologies, with the overall effect of delaying the evolution of markets, especially in the rapidly evolving and highly competitive information and communications technology industry (Blind and Thumm, 2004). In the patenting process, many organizations choose to trade off the disadvantage of eventual disclosure of the details of their technology with the guarantee of monopoly for even a limited time (Scherer and Ross, 1990).

The desire to protect IPR seems to be in direct conflict with the goal of standardization (Carlsson and Stankiewicz, 1999; Grindley, 1995; Lea and

Hall, 2004). Examining case studies, Grindley (1995) argues that the objectives of an organizational strategy for standardization are to develop a common and broadly accepted standard so as to maximize revenues, and to compete effectively once standards have been set. To maximize revenues, an organization should charge high license fees, yet in order to make its technology widely accepted, it should charge low license fees. Organizations trying to standardize their IPR-protected technology can find themselves in a real dilemma if they do not already have a large installed base, if they are not able to move others to adopt their technology, if there are compatible competing technologies, or if there are dominant players who can develop substitutable technologies.

If an IPR holder opens its technology fully, other players may jump in to take advantage of the innovative technology and the IPR holder could lose control of the technology without receiving many benefits. For example, IBM's decision to open its PC architecture quickly made it the reigning architecture for personal computers. But IBM lost control of the architecture, and subsequently lost market share to other PC manufacturers. IBM tried to regain control by introducing more advanced but proprietary technology as a standard (Grindley, 1995; West and Dedrick, 2000), but these attempts were rejected by other players who preferred to stay with the open IBM PC architecture.

On the other hand, if the IPR holder does not open its technology enough, other players will refuse to adopt it, as was the case with Sony's Beta videocassette format (Grindley, 1995). All in all, it is very challenging for an IPR holder to standardize its technologies while keeping control as an IPR holder.

As discussed, a common method for achieving the economic factors necessary for standardization is to collaborate. Organizations with IPR for their technology must make decisions about the tradeoffs involved in working with others. Lea and Hall (2004) examine the economic factors affecting an IPR holder's decision to join a standard-setting group, starting from the assumption that three factors affect any organization's decision to join a group for promulgating a standard: "(1) technological capacity and capability, (2) potential availability of network externalities, and (3) expected ability to appropriate rewards from a standard if successfully set." They find that the greater the reward expected by an IPR-holding organization, the less likely it is to join a standard-setting organization at the risk of compromising its IPR. However, they also find that developing an IPR-protected technology into a standard by oneself can make the organization vulnerable to anti-trust action.

Blind and Thumm (2004) look specifically at how the potential benefits of exercising patent rights affects the decision of patent-holding companies to join collaborations. Their findings confirm those of Lea and Hall (2004): the more patents held by an organization, the less likely it is to join a standard-setting organization. The ownership of IPR and the competitive advantage thereby accrued are more important to the organization than seeking the level of externality usually thought necessary for standardization. However, not all standard-setting collaborations require loss of control of IPR. Lemley (2002) finds that standard-setting organizations actually have diverse policies regarding the IPR of the members, ranging from mandating RAND (Reasonable and Non-Discriminatory) licensing of IPR to the group to allowing the members to have total control of their IPR.

4.4. Towards a Synthesis of the Three Strands

The above three strands of literature analyze different aspects of standards strategy in depth. Researchers have in general focused on one single aspect, and sought insight to industry-wide standards strategy by creating typologies and identifying nuances.

In this book, it is proposed to analyze organizational strategies in the ICT sector with regard to the following three aspects by creating a synthesis.

1. **Configuration of Value Network:** An organization needs to configure its value network to provide a complete product based on possible technological configurations, involving different actors that can provide necessary components or interfaces. The value network, which refers to the end-to-end network of actors that provide components, interfaces or complementary products, must be configured to deliver a complete system to end-customers (Christensen and Rosenbloom, 1995; Li and Whalley, 2002; Rudberg and Olhager, 2003). The meaning of value and value network will be discussed in detail later. This aspect is based on literature on economic factors and organizations' approaches for standardization, specifically in the decision of whether and how to collaborate with other organizations. As an organization pursues standardization, the formation of a value network can be part of a solution to the question of how to stimulate the necessary economic factors for standardization. The formation of a value network also reflects some of an organization's decisions about who to collaborate with and how. Thus, the formation of a value network can be a key component of an organization's standards strategy.

2. **Formation of Standard-Setting Organization:** An organization must decide whether it will create or join a formal or informal standard-setting organization or collaboration to standardize its technology or sustain it once standardized. This aspect comes directly from the literature on organizations' approaches for standardization, specifically in the decision of whether and how to collaborate with other organizations and indirectly related to economic factors. This is

a key aspect of an organization's standards strategy, as described by researchers (see above).

3. **Openness of IPR (Intellectual Property Rights):** Strong IPR can control a value network, but it can also reduce the availability of complementary products or compatibility of products/services, creating a less compatible product that cannot arouse the necessary economic factors. An organization must decide the extent to which it will open or close IPR as part of its standards strategy. We need to acknowledge that an actor's IPR strategy is not a simple dichotomy between being open or closed. There are in fact various types of openness, as will be explained in greater detail in the section of Aspects of Standards Strategies in Chapter 2. As shown in the literature, IPR issues in standards strategy is indirectly related to other economic factors and to organizations' approaches to standardization. We have seen that IPR becomes an important issue for organizations in deciding whether or not to join collaborations with others, because of their concern of compromising or losing control over their IPR.

The three aspects of value network, standard-setting organization, and IPR are not independent but are intimately related to each other. An organization must consider different possible options in the formation of standard-setting while at the same time plan on how to configure its value network with its and others' IPR capabilities. That is to say, an organization pursuing its standards strategy must seek a synthesis of the three aspects. In this section, these three aspects for organizational standards strategy are introduced briefly to help lay out more clearly the aspects of standards strategy that any in-depth research must account for. Now that the issues and aspects have been laid out, the challenge is to model how organizations in fact go about, successfully or not

so successfully, synthesizing the manifold issues and aspects so as to formulate their standards strategy and make adjustments as the process of standardization unfolds.

REFERENCES

Arthur, W. B. (1996). Increasing returns and the new world of business. *Harvard Business Review, 74*(4), 100–109. PMID:10158472.

Bekkers, R. (2001). *Mobile telecommunications standards: GSM, UMTS, TETRA, and ERMES.* Boston, MA: Artech House.

Besen, S. M., & Farrell, J. (1994). Choosing how to compete: Strategies and tactics in standardization. *The Journal of Economic Perspectives, 8*(2), 117–131. doi:10.1257/jep.8.2.117.

Blind, K., & Thumm, N. (2004). Interrelation between patenting and standardisation strategies: Empirical evidence and policy implications. *Research Policy, 33*(10), 1583–1598. doi:10.1016/j.respol.2004.08.007.

Boar, B. H. (1984). *Application prototyping: A requirements definition strategy for the '80s.* New York, NY: John Wiley & Sons, Inc..

Bores, C., Saurina, C., & Torres, R. (2003). Technological convergence: A strategic perspective. *Technovation, 23*(1), 1–13. doi:10.1016/S0166-4972(01)00094-3.

Carlsson, B., & Stankiewicz, R. (1991). On the nature, function and composition of technological systems. *Journal of Evolutionary Economics, 1*(2), 93–118. doi:10.1007/BF01224915.

Chandler, A. D. (1962). *Strategy and structure: Chapters in the history of the industrial enterprise.* Cambridge, MA: The MIT Press.

Chandler, A. D. (1990). *Scale and scope: The dynamics of industrial capitalism.* Cambridge, MA: Harvard University Press. doi:10.2307/3115503.

Chiesa, V., & Toletti, G. (2003). Standard-setting strategies in the multimedia sector. *International Journal of Innovation Management, 7*(3), 281–308. doi:10.1142/S1363919603000829.

Christensen, C. M., & Rosenbloom, R. S. (1995). Explaining the attacker's advantage: Technological paradigms, organizational dynamics, and the value network. *Research Policy, 24*(2), 233–257. doi:10.1016/0048-7333(93)00764-K.

Cohen, W. M., Goto, A., Nagata, A., Nelson, R. R., & Walsh, J. P. (2002). R&D spillovers, patents and the incentives to innovate in Japan and the United States. *Research Policy, 31*(8-9), 1349–1367. doi:10.1016/S0048-7333(02)00068-9.

David, P. A. (1985). Clio and the economics of QWERTY. *The American Economic Review, 75*(2), 332–337.

David, P. A. (1997). *Path dependence and the quest for historical economics: One more chorus of the ballad of QWERTY.* Oxford, UK: University of Oxford.

David, P. A., & Greenstein, S. (1990). The economics of compatibility standards: An introduction to recent research. *Economics of Innovation and New Technology, 1*, 3–41. doi:10.1080/10438599000000002.

David, P. A., & Steinmueller, W. E. (1994). Economics of compatibility standards and competition in telecommunication networks. *Information Economics and Policy, 6*(3-4), 217–241. doi:10.1016/0167-6245(94)90003-5.

Farrell, J., & Saloner, G. (1986). Installed base and compatibility: Innovation, product preannouncements, and predation. *The American Economic Review, 76*(5), 940–955.

Faulkner, D., & Campbell, A. (2003). Introduction to volume 1: Competitive strategy through different lenses. In D. Faulkner & A. Campbell (Eds.), The Oxford Handbook of Strategy - Volume 1: A Strategy Overview and Competitive Strategy (pp. 1-17). New York: Oxford University Press.

Ferguson, R. B. (2006). RFID standard battle rages: HF or UHF? *eWEEK.com.* Retrieved from http://www.eweek.com

Ferguson, R. B. (2006). Wal-Mart's new CIO Says he'll back RFID. *eWEEK.com.* Retrieved from http://www.eweek.com

Fifield, P. (1992). *Marketing strategy.* Oxford, UK: Butterworth-Heinemann Ltd..

Fine, C. H. (1998). *Clockspeed: Winning industry control in the age of temporary advantage.* Reading, MA: Perseus Books.

Frohlich, M. T., & Westbrook, R. (2002). Demand chain management in manufacturing and services: Web-based integration, drivers and performance. *Journal of Operations Management, 20*(6), 729–745. doi:10.1016/S0272-6963(02)00037-2.

Gambardella, A., & Torrisi, S. (1998). Does technological convergence imply convergence in markets? Evidence from the electronics industry. *Research Policy, 27*(5), 445–463. doi:10.1016/S0048-7333(98)00062-6.

Goold, M., & Luchs, K. (2003). Why diversify? Four decades of management thinking. In D. Faulkner & A. Campbell (Eds.), The Oxford Handbook of Strategy - Volume 2: Corporate Strategy (pp. 17-42). New York: Oxford University Press.

Grindley, P. (1995). *Standards, strategy, and policy: Cases and stories.* Oxford, UK: The University Press. doi:10.1093/acprof:oso/9780198288077.001.0001.

Hilton, G. W. (1990). *American narrow gauge railroads*. Stanford, CA: Stanford University Press.

Katz, M. L., & Shapiro, C. (1985). Network externalities, competition, and compatibility. *The American Economic Review, 75*(3), 424–440.

Katz, M. L., & Shapiro, C. (1994). Systems competition and network effects. *The Journal of Economic Perspectives, 8*(2), 93–115. doi:10.1257/jep.8.2.93.

Kay, J., McKiernan, P., & Faulkner, D. (2003). The history of strategy and some thoughts about the future. In D. Faulkner & A. Campbell (Eds.), The Oxford Handbook of Strategy - Volume I: A Strategy Overview and Competitive Strategy (pp. 21-26). New York: Oxford University Press.

Lea, G., & Hall, P. (2004). Standards and intellectual property rights: An economic and legal perspective. *Information Economics and Policy, 16*(1), 67–89. doi:10.1016/j.infoecopol.2003.09.005.

Lee, J. Y., & Chan, K. C. (2003). Assessing the operations innovation bandwagon effect: A market perspective on the returns. *Journal of Managerial Issues, 15*(1), 97–105.

Lemley, M. A. (2002). Intellectual property rights and standard-setting organizations. *California Law Review, 90*(6), 1889–1980. doi:10.2307/3481437.

Li, F., & Whalley, J. (2002). Deconstruction of the telecommunications industry: From value chains to value networks. *Telecommunications Policy, 26*(9-10), 451–472. doi:10.1016/S0308-5961(02)00056-3.

Liebowitz, S. J., & Margolis, S. E. (1990). The fable of the keys. *The Journal of Law & Economics, 33*(1), 1–25. doi:10.1086/467198.

Liebowitz, S. J., & Margolis, S. E. (1994). Network externality: An uncommon tragedy. *The Journal of Economic Perspectives, 8*(2), 133–150. doi:10.1257/jep.8.2.133.

McGee, J., & Sammut Bonnici, T. A. (2002). Network industries in the new economy. *European Business Journal, 14*(3), 116–132.

McLaughlin, C. P., Pannesi, R. T., & Kathuria, N. (1991). The different operations strategy planning process for service operations. *International Journal of Operations & Production Management, 11*(3), 63–76. doi:10.1108/EUM0000000001268.

Merriam-Webster. (1998). *Webster's ninth new collegiate dictionary*. Springfield, MA: Merriam-Webster Inc..

Miller, A., & Dess, G. G. (1996). *Strategic management*. New York: McGraw-Hill.

OECD. (2002). *Measuring the information economy*. Paris: OECD.

OECD. (2006a). *Information technology outlook*. Paris: OECD.

OECD. (2006b). *Science and technology*. Paris: OECD.

Oshri, I., & Weeber, C. (2006). Cooperation and competition standards-setting activities in the digitization era: The case of wireless information devices. *Technology Analysis and Strategic Management, 18*(2), 265–283. doi:10.1080/09537320600624196.

Powell, W. W., & Snellman, K. (2004). The knowledge economy. *Annual Review of Sociology, 30*, 199–220. doi:10.1146/annurev.soc.29.010202.100037.

Puffert, D. (2000). The standardization of track gauge on North American railways, 1830-1890. *The Journal of Economic History, 60*(4), 933–960.

Puffert, D. (2002). Path dependence in spatial networks: The standardization of railway track gauge. *Explorations in Economic History, 39*(3), 282–314. doi:10.1006/exeh.2002.0786.

Rivette, K. G., & Kline, D. (2000). Discovering new value in intellectual property. *Harvard Business Review, 78*(1), 54–66.

Rudberg, M., & Olhager, J. (2003). Manufacturing networks and supply chains: An operations strategy perspective. *Omega, 31*(1), 29–39. doi:10.1016/S0305-0483(02)00063-4.

Sammut-Bonnici, T., & McGee, J. (2002). Network strategies for the new economy. *European Business Journal, 14*(4), 174–185.

Scherer, F. M., & Ross, D. (1990). *Industrial market structure and economic performance.* Boston, MA: Houghton Mifflin.

Schilling, M. A. (2002). Technology success and failure in winner-take-all markets: The impact of learning orientation, timing, and network externalities. *Academy of Management Journal, 45*(2), 387–398. doi:10.2307/3069353.

Shapiro, S., Richards, B., Rinow, M., & Schoechle, T. (2001). Hybrid standards setting solutions for today's convergent telecommunications market. In *Proceedings: 2nd IEEE Conference on Standardization and Innovation in Information Technology* (pp. 348-351). Boulder, CO: IEEE.

Sullivan, L. (2004, December 13). Team of the year. *InformationWeek*.

Tassey, G. (2000). Standardization in technology-based markets. *Research Policy, 29*(4-5), 587–602. doi:10.1016/S0048-7333(99)00091-8.

Tripathi, M. (2006). Transforming India into a knowledge economy through information communication technologies—Current developments. *The International Information & Library Review, 38*, 139–146. doi:10.1016/j.iilr.2006.06.007.

Tsikriktsis, N., Lanzolla, G., & Frohlich, M. (2004). Adoption of e-processes by service firms: An empirical study of antecedents. *Production and Operations Management, 13*(3), 216–229. doi:10.1111/j.1937-5956.2004.tb00507.x.

van Wegberg, M. (2004). Standardization process of systems technologies: Creating a balance between competition and cooperation. *Technology Analysis and Strategic Management, 16*(4), 457–478. doi:10.1080/0953732042000295784.

West, J., & Dedrick, J. (2000). Innovation and control in standards architectures: The rise and fall of Japan's PC-98. *Information Systems Research, 11*(2), 197–216. doi:10.1287/isre.11.2.197.11778.

Yoffie, D. B. (1996). Competing in the age of digital convergence. *California Management Review, 38*(4), 31–53. doi:10.2307/41165853.

Chapter 2
Developing a Theoretical Model

EXECUTIVE SUMMARY

Much of the literature on standards strategy focuses on classifying the various observed strategies. However, what is not clearly evident from the literature is an account of why and how organizations choose their strategies. By what logic does any company decide how to promote the necessary economic factors to spread its technology, how to create a value network built upon that technology, whether to work with collaborators or go it alone, and whether to close or open IPR? What influences these decisions? What mix of elements in a company's situation will prompt it to make one choice rather than another? To go beyond a mere listing or categorization of the types of standards strategies, but try to understand why and how organizations choose certain strategies, a theoretical model to analyze the organizational process for deciding upon standards strategy is needed. In this chapter, by reviewing existing theories that researchers have used in the literature on standard-setting processes, a theoretical model for the purpose of this study is proposed.

1. EXISTING THEORIES

In the research on technology standardization, three main theories or approaches have been employed. Early studies used economic theories such as economies of scale and network externalities to identify economic factors that promote standardization (e.g. David and Steinmueller, 1994; Katz and Shapiro, 1985; Liebowitz and Margolis, 1994). Since then, some researchers have adopted Game Theory to analyze organizations' strategic

approaches to setting standards (e.g. Besen and Farrell, 1994; Chiesa and Toletti, 2003). Other researchers have used the model of Actor Network Theory, or ANT, to describe and interpret the involvement and interaction of the many actors in the actual process of technology standardization (e.g. Graham et al., 1995; Hanseth et al., 2006; Markus et al., 2006).

Economic theories such as economies of scale and network externality have been adopted widely to identify economic factors in standardization. These theories help to clarify the economic impact of standardization from the perspective of the

DOI: 10.4018/978-1-4666-4074-0.ch002

entire industry. From the perspective of an individual organization, however, the challenge is to figure out its strategy to influence the economic factors and to shape the industry standards to its own benefit.

Researchers have suggested one way to stimulate economic factors by identifying different types of standard-setting processes according to whether organizations collaborate with others in a formal or informal way (e.g. Besen and Farrell, 1994; Chiesa and Toletti, 2003). The basic theory used for modeling these types of standard-setting processes and the strategic decisions involved is Game Theory. The theory of games, developed by mathematician John von Neumann and economist Oskar Morgenstern as a strain of applied mathematics and economics, has been used by researchers to study various competitive and cooperative situations where actors strive to pursue strategies that would maximize their payoffs (Von Neumann and Morgenstern, 1953).

Game theory studies on standards contribute to our understanding through the identification of the different types of collaborations and standard-setting processes or collaborations, and through their analysis of the relationship between an organization's situation (for example, its market position) and its circumscribed choice of standard-setting processes. However, these game theory models make drastic simplifying assumptions about the structure and process of technology standardization. For example, many models assume that there are only two competing organizations and that they have similar market and technology positions. As a result of this simplifying approach to modeling, the kind of standards strategies considered in the model is restricted to the making of the simple decision on choosing this or that standard process, and this or that form of collaboration.

In a practical business environment, the topology of the standard setting game is rather convoluted. First of all, there are usually a lot more than two competing companies. Each of the various types of players—users, vendors, technologists,

regulators, etc.—push their own agendas and hedge their bets by forming multiple alliances and coalitions. Third, the form and process that the industry ultimately elects to settle technology standards are determined endogenously through the actual playing of the standard-setting games (and not exogenously given as strategic choices). All in all, we need a richer framework of analysis in order to account for the "topology" of technology configuration, value network, and distribution of IPR so that one might actually understand how order (i.e. an agreed standard) emerges out of chaos of the conflicting interests and the competing maneuvers of the many players.

Recently more researchers have recognized the need to account for the complexity of the involvement of many actors in the process of technology standardization. Some researchers have adopted Actor Network Theory (ANT) to model the conflicting goals and purposes of the actors and how they interact. ANT was initially developed at the Ecole des Mines in Paris (Callon, 1986; Latour, 1987) to study the sociology of science. Other researchers later applied the theory to study technology development (Latour, 1996). It is now being used broadly in sociology, feminist studies, organizational studies, and other interdisciplinary areas such as science and technology evolutions (e.g. Gao, 2005; Lea et al., 1995; Sarker et al., 2006). According to Kaghan and Bowker (2001), an actor network is defined as "any collection of human, non-human, and hybrid (human/non-human) actors who jointly participate in some organized (and identifiable) collective activity in some fashion for some period of time" (p. 258). Once an actor network takes shape, it operates as a whole for as long as the shared goal of the actors remains the same. The network is not always intrinsically coherent and can evolve as conflicts and interests among actors develop or intensify. For this reason, the network can change over time and even fall apart.

For example, Lea et al. (1995) apply ANT to understand the context and the development of electronic communications within one organization. They conducted a case study on the interactive process with which a Europe-wide consulting firm built an e-communication system that adapts to the different and sometimes contradictory interests among various actors. Gao (2005) uses ANT to identify the interactive process through which the Chinese government's strategies for transforming the telecommunications market were realized. He models the Chinese telecommunications market as a non-human actor, and the public, the state, and telecommunications service providers as human actors. This study analyzes how these actors collectively shaped the context of the Chinese telecommunications market, and how the concrete content of the Chinese government's strategies took form over time.

The potential usefulness of ANT is now more broadly known among researchers studying information and communications technology, including technology standardization (e.g. Graham et al., 1995; Markus et al., 2006; Hanseth et al., 2006). For example, Walsham and Sahay (1999) apply ANT to analyze the development and usage of Geographical Information Systems for district-level administration in India. Another example is Rohracher (2003), who uses ANT to understand the role of end-users in social shaping of environmental technologies.

In the technology standardization field, Graham et al. (1995) use ANT to analyze the actors' internal tensions and contradictions in the EDIFACT (Electronic Data Interchange for Administration Commerce and Trade) message development process, which was part of the informal process in developing the international standards for EDI. They argue that ANT is a highly appropriate framework for understanding the many actors' motivations and actions in EDI standardization. More recently, Markus et al. (2006) present a case study of industry-wide information systems standardization in the U.S. residential mortgage industry. They use ANT to model the heterogeneous interests of various actors such as software

vendors, mortgage bankers, and credit reporting companies, and how these interests interactively shape the process of developing and diffusing standards for setting up information systems in the mortgage industry. Their findings emphasize the necessity of accommodating the interests of the heterogeneous participants—for example, the user groups' participation in standards development and their commitment to adopt the standard—and the necessity of fitting the technical content of the standard with the heterogeneous needs of the participants.

One advantage of using ANT to study technology standardization is that it underscores various actors' involvements and their conflicting strategies, whose interaction can lead to unexpected or innovative results in standardization (Hanseth et al., 2006). ANT also provides a more practical view of the business environment in technology standardization by incorporating any number and variety of actors and their strategies. However, ANT lacks the capacity to address some important aspects of standardization since it treats human and non-human actors on the same level. Obviously, human actors have intentions and their intentions can vary within different situations, while non-human actors do not have conscious intentions. Therefore, ANT is not set up to explain why and how certain phenomena are manipulated intentionally to arise; ANT simply aims to help describe what actually happens (Gao, 2005).

Economic theory, Game Theory, and ANT (Actor Network Theory) provide three different approaches to research on technology standards and strategies by highlighting different aspects of standardization. Economic theory emphasizes the economic factors in standardization. Game Theory proposes a way to understand the relationship between an organization's choice of standard-setting process and its situation, under some simplifying theoretical assumptions. With a practical view of the business environment, ANT describes the many actors' involvement and the interaction of their conflicting strategies.

Although each of these theories provides an approach to understand one or two aspects of standards strategy, we need an integrated framework to analyze organizational standards strategies that can afford a broader, holistic view and which takes into account the conditions of the real-world business environment. With such a framework, we can answer questions such as what organizational standards strategy can be; what situations make organizations decide on a certain standards strategy; why organizations in similar situations develop different standards strategies; and what the consequences are of certain organizational standards strategies.

2. THEORETICAL DEVELOPMENT

To do research on organizational standards strategies from an integrated perspective, it is critical to have an appropriate theoretical framework. Because of ANT's practical view of the business environment, it is argued that among the three main theoretical frameworks used by other researchers, ANT provides the best starting point. Economic theory is limited to explaining the factors that contribute to the setting of standards strategy, but cannot be used to explain the motivations of organizations, or address the larger picture of how they interact. Game theory relies on simplifying assumptions that limit its usefulness for analyzing real life situations.

ANT is useful because it does take into account the complexity of real life and can model the interactions of organizations. However, ANT cannot explain the context in which actors make decisions – for example, why do certain networks emerge and how do they emerge? In order to answer these broader questions, we need to be able to analyze industry-wide dynamics and describe the larger picture of how the many organizations interact with one another. For this purpose, the theory of Self-organized Complexity will be drawn.

Psychiatrist and engineer W. Ross Ashby first introduced the term "self-organizing" in his paper "Principles of the Self-organizing Dynamic System" in 1947. Since then, the concept of Self-organized Complexity, which is also called Complexity Theory or Self-Organization Theory, has been adopted widely in the physical, biological and social sciences as well as in economics and finance, to understand the arising of structures and patterns in complex phenomena (e.g. Estep, 2006; Goldbaum, 2006; Krugman, 1996; Stanley et al., 2002; Turcotte et al., 2002).

In the natural sciences, the concept of self-organization was developed from observations of phenomena in biology and physics in which systems with multiple interacting components were seen to form certain regular structures or patterns without any apparent external motivating or regulating forces (Haken, 2006; Smith and Graetz, 2006). Physicist Haken (2006) defines *self-organization* as a characteristic of a system "if it acquires a spatial, temporal or functional structure or pattern without *specific* interferences from outside" (p. 11, emphasis added). He emphasizes the continuous flow of non-specific interferences that increases the complexity of the system, because it appears that self-organization occurs only when the environment is complex and fluid.

For example, biologists use Self-organized Complexity to explain the flying formation of geese when they migrate. Migration is a very complex situation for geese: countless birds with limited cognitive power and communication ability fly together for long distances, most of them not knowing where they are heading. However, they are able to self-organize into the shape of a "V" with a single lead goose in front and two lines of geese streaming out behind (Camazine et al., 2003). Another example is the self-organized complexity of the Internet. The Internet is a hugely extensive connection of myriad nodes using various network links and media. It is a very complicated network without any predetermined organization or structure. Each node does not have any comprehension of the whole network but only communicates with neighboring nodes. The nodes, however, by adapting to local conditions when they pass data and communicate with each other, somehow collectively fall into certain self-organized patterns (Turcotte and Rundle, 2002). In these and other examples, self-organized complexity refers to the fact that, instead of chaos, coherent patterns somehow do arise out of complex unplanned and largely local interactions.

The concept of self-organization is adapted from the physical to the social sciences with some modifications. Social scientist point out that in the social sciences, the boundary between outside or inside is not clear in many cases and the environment cannot be controlled to the extent possible as in physics experiments (Anderson, 1999; Smith and Graetz, 2006). It is necessary to understand "complexity" first before reaching a definition of self-organization suitable for social science research.

Researchers have argued that complexity prevails when a large number of parts have many interactions between them (Simon, 1996) or when there is a set of interdependent parts, which together build up to a whole (Thompson, 1967). Morel and Ramanujam (1999) indicate that while it is difficult to define complexity, it is easy to recognize. Instead of providing a definition, they state what they consider as the readily identifiable characteristics of complexity: (1) having a large number of interacting elements, and (2) having emergent properties, i.e. the appearance of patterns as a result of the collective behavior of interacting elements. Other researchers have added more characteristics since then. For instance, E Cunha and Da Cunha (2006) add that the interactions between elements are local, recursive and non-linear; Smith and Graetz (2006) indicate that the emergent behavior does not necessarily follow predictably from underlying causes.

These definitions and characteristics of complexity provide us with the basic elements of self-organized complexity: (1) the presence of many actors, (2) interactions among actors that are non-linear, and (3) configurations that emerge based on those interactions. In the case where there are many actors and where the manifold interactions among them fail to emerge any orderly configurations or patterns, we would call such a state of affairs disorder or chaos (cf. Morel and Ramanujam, 1999).

Given these definitions and characteristics of "complexity," Anderson (1999) defines self-organization as "systems that consist of independent actors whose interactions are governed by a system of recursively applied rules that naturally generate stable structure. They self-organize; pattern and regularity emerge without the intervention of a central controller" (p. 221). His definition is based on the study of Drazin and Sandelands (1992) in the social sciences. As pointed out, the boundary between outside and inside is ambiguous in social sciences, thus, this definition refers to "interference of a central controller" instead of "specific interference from outside" (e.g. Anderson, 1999; Drazin and Sandelands, 1992).

One example in the business field is offered by Smith and Graetz (2006), who apply the concept of Self-organized Complexity in organizational management to provide managers a way to encourage activities among employees to produce innovative ideas. The structure created to do this might seem contradictory to the definition of self-organization, but they argue that management guidance is not necessarily the opposite of self-organization, but can go together by the development of a soft-system (which refers to loose-structure). Through this soft-system, the organization can promote the right kind of local actions that encourage self-organization. A controller, in this case a manager, cannot manipulate self-organization directly, but (s)he can promote the necessary conditions for self-organization.

Another example is Benbya and McKelvey's (2006) application of Self-organized Complexity theory to IS (Information Systems) alignment. IS alignment includes alignments between business strategy and IS strategy, between organizational structure and IS structure, and between users' needs and IS infrastructure. Because of the complex mutual influences and the co-evolutionary nature of the relationships and interactions between IS and other organizational areas, Benbya and McKelvey (2006) emphasize the need for a different approach to study IS alignment that differs from the traditional view which has "a tendency to focus upon simple cause-effect deterministic logic" (p. 284). They apply the idea of self-organized complexity and co-evolution to explain how IS alignment actually emerges, instead of seeing it as serendipity.

Drawing on this line of research, this study will adopt the following definition of self-organization:

Self-organization occurs when a configuration or pattern emerges from the interaction of various independent actors over time, without the intervention of a central controller (Drazin and Sandelands, 1992; Anderson, 1999).

Why is this theory appropriate for analyzing technology standardization? Technology standardization is a very complex phenomenon (Grindley, 1995; Hanseth et al., 2006). Many organizations have their own interests and strategies, with which they try to influence others (Grindley, 1995; Seo and Mak, 2010). An organization's interests and focus change over time, depending on how their (non-linear) interactions with other organizations work out, and how its situation evolves (Anderson, 1999; E Cunha and Da Cunha, 2006). In general the business environment does not degenerate into chaos. Somehow, organizations find a way to set workable technology standards, sometimes through voluntary standard-setting collaboration, sometime through the emergence of a market-dominating technology. Clearly, the process of

technology standardization meets the characteristics of self-organized complexity. Therefore, it is believed that Self-organized Complexity is the right perspective from which to develop a framework for analyzing the process of technology standardizations.

Therefore, it is proposed to integrate the ANT with Self-organized Complexity to create a theoretical framework that can support a holistic analysis of organizational standards strategy. The Self-organized Complexity perspective provides an approach to analyze the overall phenomenon of technology standardization (Anderson, 1999), because it recognizes the existence of many actors and their non-linear interactions that self-organize into patterns. The ANT perspective helps us trace and understand the dynamic relationships among actors. Therefore, the combination of these two perspectives should allow us to explain not only the formulation of standards strategies of individual organizations, but also how these strategies interact and ultimately, through non-linear feed-forward and feed-back loops, "produce" the observed technology standards and industry structure.

This integrated framework is schematically represented by the following the Self-Organized Complexity Unfolding Model (Figure 1), based on the ideas of Mak (2006).

The Self-organized Complexity Unfolding Model refers to the following dynamics: (1) the existing self-organized configurations or patterns in the industry affect each individual actor's situation; (2) each actor interprets its situation through its orientation, values, and goals; (3) each actor creates strategies based on its interpretation and takes actions according to these strategies; (4) these actions of individual organizations interact to create or shape value networks; (5) the workings of these value networks "feed forward" to affect and modify the configuration of the industry; and (6) the evolving industry configuration in turn "feed back" to affect and modify the situations of the organizations. This continuous process of feed forward and feedback between

Figure 1. Self-organized complexity unfolding model

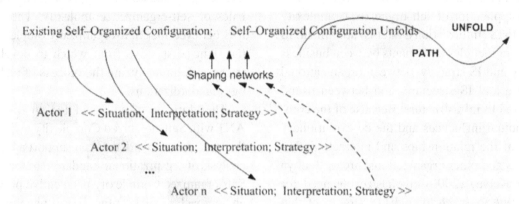

organizational strategies and actions and the industry configuration of value networks over time maps out a path through which the industry structure emerges, unfolds, self-organizes, and then becomes entrenched.

This Self-organized Complexity Unfolding Model is a very general conceptual model. What follows is an explanation of how this model may be applied to the particular study of organizational strategies for ICT standards.

Actors in ICT standardization include technology-creating organizations that want to standardize the technology they have developed, and technology-adopting organizations (e.g. the complementary product manufacturers), end-customers that accept technology-embedded products or services, and governments that can influence the technology standardization directly or indirectly with regulations (Blackman, 1998; Bores et al., 2003; Shell, 2004). In this book, only human actors are included, because we are only considering actors that have some will to create strategies, while non-human actors such as technology are passive objects that do not have the will or the ability to interact with other actors intentionally.

The terms "actor" and "organization" are used equivalently in this book, so organization can refer not only to a company or corporation but also to a government or a voluntary standard-setting

organization formed by various companies or governments.

However, ownership of technology IPR and technical competence are significant parts of actors' situations in standardization battles. An actor's posture with respect to technology influences its interactions with other actors, its participation in value networks, and by extension the competition among the value networks with different postures in technologies. Therefore, technology itself should be considered as a major constituent of the industrial configuration.

Existing self-organized configurations or patterns of ICT refer to the contextual industry environment in which actors operate. For example, the early PC market was fragmented, meaning that PCs released by different companies were not compatible with each other. That was the existing configuration which affected all actors at the time.

Although there is one overall industry-wide context relevant to all actors in the industry, the concrete and local contextual environment of each individual actor is different, manifested as its particular market position, resources and capabilities which together govern its interactions with other actors (Anderson, 1999). This "local" context is the actor's "situation." Since all actors' actions and decisions are based on their sense on local environments, an actor is necessarily sensitive to its situation (E Cunha and Da Cunha, 2006).

"Interpretation" refers to the way each actor recognizes and perceives its situation. The situation refers to the objectively observable characteristics of an actor's environment, while interpretation of the situation is a cognitive process which is highly subjective (Anderson, 1999). Because an objective situation might be subjectively interpreted very differently, two actors could pursue very different strategies even though their situations may be similar.

Each actor tries to carry out its strategies through various tactics such as developing new technologies, expanding its market for existing products, introducing new products, and forming strategic alliances with others. These tactics form the basis of interactions between actors, and manifest as activities such as competition in the marketplace, cooperative teamwork to develop and deliver products together, etc. (Anderson, 1999).

Extensive and varied interactions between actors contribute to the complexity of the industry, and at the same time serve as the basis of self-organizing (Morel and Ramanujam, 1999). From the perspective of ANT, interactions are important because networks are formed and shaped through them (Hanseth et al., 2006). From the perspective of Self-organized Complexity, networks are important because they are the organized form through which actors' activities may generate values – witness the workings of the networks of supply chains, of trade alliances, and of standard setting organizations.

These value networks are not independent of each other. Two value networks can compete; one network may grow at the expense of other networks by inducing actors to switch over. This is so with respect to market competition; likewise for gaining support for competing technology standards; likewise for forming trade and political alliances. Other than relative strength, the interactions among the various networks are also affected by forces such as the bandwagon effect and network externalities.

The interactions between actors feed forward and feed back within the context of a value network. At a higher level, the interactions between value networks feed forward and feed back within the context of the self-organizing industrial configuration (cf. Anderson, 1999). Obviously, it is nontrivial to try to sort out and analyze explicitly the workings of such multi-level complex dynamic systems. Indeed, it is highly questionable whether any generally valid statements may be made about how such systems actually self-organize.

To proceed further, with the goal of explaining how technology standards actually arise, we must add more concrete structure and detail to the Self-organized Complexity Unfolding Model for technology standardization. This will be done in the following section.

3. FRAMEWORK FOR ANALYZING STANDARDS STRATEGY

The Self-organized Complexity Unfolding Model provides a general approach to understanding the complexity of the standards strategies of organizations and their interactions that unfold over time to create an industry's business configurations. To systematically trace, model, understand, and possibly predict the unfolding of self-organized configurations, one must begin by cutting time into slices and taking snapshots of organizations' situations, how they interpret their situations, and their development of strategies and tactics. Then, by drawing connections among these snapshots judiciously, one may piece together a *picture* of how and why the industry configuration evolves and unfolds over time, as influenced by the organizations' standards strategies.

Suppose one adopts the modeling approach of the usual discrete dynamic models. Then one would designate certain discrete points of time as, Time 1 (T1), Time 2 (T2), …, Time n (Tn), and the industry configuration would be modeled as some vector of states at these times. Then (cf. Anderson, 1999), how an existing self-organized configuration influences an organization's situation at time Ti, and how an organizations interprets its situation at time Ti would be modeled through certain complex functions – standing for the strategy of the organization – that map certain sub-vectors of the states to certain actions of the organization. These "action functions" of the organization would together determine (via some modeled interaction mechanism) the future vector of states in time T(i+1). As such a dynamic model iterates through times T1, T2, T3 and so on, the time series of the state vector would then be taken as a model of the unfolding of the industry configuration.

One could force the Self-organized Unfolding Complexity Model outlined earlier to take on the specific form of a traditional discrete dynamic model. However, doing so would preclude any possibility of discovering and understanding the self-emergence of industry configurations. This is because a formal model must specify ahead of time both the "state space" and the legitimate "interaction mechanism." For cases where strategic intent is not relevant—e.g. the interactions of physical particles, of genes and cells, and of simple animals (e.g. geese flying in formation)—such an approach to modeling is doable and indeed it is through such an approach that researchers first identified and generated understanding of the phenomenon of self-organization and emergence.

But such a formal and restrictive approach would hardly be appropriate for the research at hand which is deeply concerned with organizations' strategic intents and creative maneuverings. This claim will be justified by the details of the model to be laid out presently. Here, it will be brought up the point that the interactions among the organizations are by and large through the co-opetition of the value networks they self-organize. Therefore, the "interaction mechanism" is, so to speak, generated endogenously and cannot be modeled or prescribed ahead of time. A similar point

may be made with respect to the characterization of configurations (i.e. the state space).

However, if one is to investigate the unfolding of phenomena over time, one can hardly avoid taking time slices and tracing through the iteration of the actors' actions over time. Here, it is explained how time slices are to be taken and used in this book, and how iteration of action over time is to be conceived.

First, the time points Ti's are not fixed or predetermined. In the unfolding of technology and industry configuration, there are natural rhythms and life cycles composed of phases of exploration, initial development, take-off, maturation, and demise. And each organization, within the context of the industry life cycles, has its own specific life cycles. By time points, it is really referred to certain epochs (time periods) suitable for the discernment of the proceedings of the various big and small life cycles.

At any time Ti, the term "configuration" refers not only to what is going on at Ti, but because of the unfolding life cycles reach back to T(i-1) or further, and project forward to T(i+1) or beyond (see Figure 2). The term "situation" of an organization should be understood likewise. Most importantly, when an organization "interprets" its situation at time Ti, it is understood that the organization attempts to take stock of its cumulative experience up to Ti, and attempts to see how the ongoing projects, value networks, and life cycles will project forward to time T(i+1) and beyond.

When a time slice is cut in the Self-organized Complexity Unfolding Model, one obtains a cross-sectional framework for organizational standards strategy (Figure 3). Specifically, each organization is seen (or rather dissected) as composed of three main parts: the organization's situation, its interpretation of the situation, and its standards strategy.

Each of these three main parts has several elements or aspects. An overview of the modeling of these aspects is given here, with the details of the model to follow.

Figure 2. Dissecting the self-organized complexity unfolding model

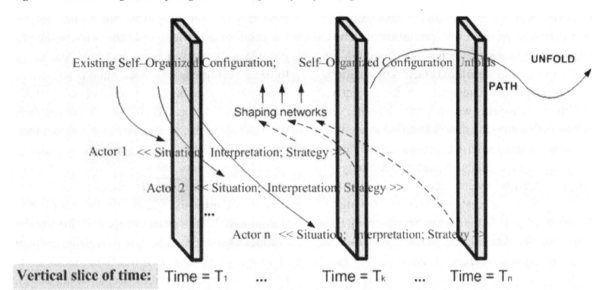

Figure 3. Cross-sectional framework for organizational standards strategy

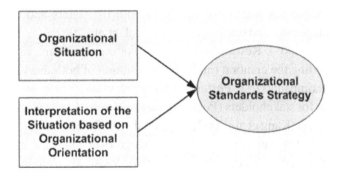

An organization's situation with respect to ICT standards is characterized by the following six fundamental elements: (A) the organization's capabilities to meet market needs and opportunities, (B) the availability of complementary products or compatibility of products/services in the market, (C) the innovativeness of the technology involved, (D) the position of the organization in the market, (E) the availability of alternative or substitutable technologies in the market, and (F) the characteristics of intellectual property rights regarding the technology involved.

Secondary, organizational standards strategy is modeled as having three main aspects: (1)

shaping the configuration of value networks, (2) promoting or obstructing various standard-setting endeavors, and (3) leveraging on the openness of IPR (Intellectual Property Rights).

Thirdly, an organization's interpretation of its situation is taken to be how it addresses the fundamental question of "How do I find myself?" The modeling of such interpretations points to the integration of two aspects: One is the subjective aspect of the organization's orientation, beliefs, and psychological/cultural make-up. The other is the objective aspect of how the organization fits into the industry as a whole, how it fits as one player into the game of standardization, and how it as a node fits into the unfolding value networks. As outsiders, we do not have access to organizations' processes of interpretation. Thus, we shall take an organization's "interpretation" as its observed response to the six factors of situation, and its manifested integration of its subjective and objective aspects which lead to its observed strategy regarding the three identified aspects.

Actually, this model of Cross-sectional Framework of Organizational Standards Strategy is simply a straightforward accounting of how the necessity for value creation drives the striving of ICT organizations in the current environment. Before giving the details of the model as promised, a few comments on value creation seem appropriate.

The concept of value creation and capture is the foundation of all theories of organizational strategy related to value networks – organizational behaviors are not random activities but are geared towards value creation. Value has long been a central concept in the management and organization literature (Lepak et al., 2007). Researchers have focused on understanding the creation (and destruction) of values for companies (e.g. Porter, 1985; Sirmon et al., 2007), for stakeholders (Post et al., 2002) and for customers (Kang et al., 2007; Priem, 2007). In this book, this perspective of value creation is simply adopted. It is employed as the driving force in organizational standards strategies to model how the endeavors of ICT value creation are carried out through organizations' interpretation of situation and their pursuit of standards strategy.

The concept of value is quite broad. Researchers have defined value and studied the meaning of value creation and capture at various levels – individual, organization, and society. Value at the organizational level is focused because organizational standards strategy is the focus of this book.

Bowman and Ambrosini (2000) introduce two types of value: One is use value, which refers to the perceived or realized benefits users derive from how goods and services meet their needs.

The other is exchange value, which refers to the monetary amount that users are willing to pay for goods and services and sellers are willing to provide at any given point in time. According to these definitions of value, value is subjective depending on the buyers or sellers (Lepak et al., 2007). From the perspective of companies, they need to have an exchange value that is higher than the actual cost of producing their goods, but still lower than the use value for customers in order to attract buyers. From the customers' point of view, they consider the use value first and tend to pay the amount that the sellers suggest, as long as the product or service can create new value to them or its exchange value is lower than that of alternative products or services (Lepak et al., 2007). Therefore, to understand the value of goods, it should be observed and analyzed in the context of mutual negotiation of the meanings of goods to both the sellers and buyers (Amabile, 1996; Lepak et al., 2007).

For organizations, value creation refers to the generation of both use value and exchange value (Lepak et al., 2007). Value capture refers to the actions geared towards sustaining value (Lepak et al., 2007). This is pertinent because value can change over time; for instance, exchange value can decrease when competitors introduce alternative products or services that provide similar use value with lower exchange value to customers (Lepak et al., 2007). Although much research has concentrated on increasing exchange value by developing innovative products or services without necessarily considering the increase in use value for customers, it is important for organizations to create, increase and capture use value for customers, because customers do not evaluate higher exchange value without evaluating use value as well (Kang et al, 2007; Priem, 2007).

The recent research on value creation and capture offers two important points. One is that it is necessary to include end-customers as one actor group, because of the importance of customers' use value for organizations (Kang et al, 2007; Priem, 2007). The other is that the elements or aspects that are necessary to create and capture value are the key to understanding and analyzing organizational situations and strategies because only through them can an organization gain and sustain competitive advantage (Bowman and Ambrosini, 2000; Slater, 1997; Woodruff, 1997).

For this reason, the elements and aspects that factor in specifying the organizational situations, interpretations, and standards strategy are modeled in detail. The following three sections will make clear how these elements and aspects play their role in value creation and capture.

3.1. Elements of Organizational Situation

A review of the literature helps highlight the most important elements that make up an actor's situation with regard to setting standards, including both internal and external factors. These are: (A) an organization's capabilities to meet market needs and opportunities (Besen and Farrell, 1994; Kang et al., 2007; Priem, 2007); (B) the availability of complementary products or compatibility of products/services in the market (Rosen, 1994; Chiesa and Toletti, 2003); (C) the innovativeness of the technology involved (Christensen, 1997; Seo and Lee, 2007); (D) the position of the organization in the market (Besen and Farrell, 1994; Chiesa and Toletti, 2003; van Wegberg, 2004); (E) the availability of alternative or substitutable technologies in the market (van Wegberg, 2004); and (F) the characteristics of intellectual property rights regarding the technology involved (Blind and Thumm, 2004; Lea and Hall, 2004; Shell, 2004). Note that these elements are all factors relevant to the organizations' possibilities to create and capture value.

The relevance of each of the above six elements is grounded in different theories. For example, the two elements—"an organization's capabilities to meet market needs and opportunities" and "the characteristics of intellectual property rights"—are grounded in the Resource-Based View (RBV) of the firm. The RBV theory argues that resources are asymmetrically distributed among firms, so thus the firms that possess valuable, rare, imperfectly imitable, and not equivalently substitutable resources may gain sustained competitive advantage over those that do not (Barney, 1991; Grant, 1991; Penrose, 1959; Wernerfelt, 1984). In this sense, an actor that has more capabilities to meet market needs and opportunities, and has more intellectual property rights for essential technologies is in a situation where it has strategic advantages. It should therefore leverage to develop a winning standards strategy. In contrast, the relevant element (D)—the position of the organization in the market—is grounded in Porter's competitive forces model popular in the 1980s (Porter, 1980). The competitive advantage of an organization in setting standards also come from its market position, more specifically, from its relative power position within various value networks.

Before discussing these six elements in detail, three important methodological points must be made:

- First, the objectively existing situation must be distinguished from the organization's subjective interpretation of it. Indeed, an organization might misread its situation—failing to properly account for one or more of the factors—and thereby fail to leverage its objectively strong position to gain or sustain its competitive advantages. For example, legendary CEO Andrew Grove of Intel has openly admitted that Intel did not appreciate the "tidal wave" of the emergence of the Internet and therefore did not leverage its dominance in PC hardware, essentially yielding the emerging network hardware market to new players such as Cisco.

- Second, as pointed out in the earlier introduction of the Self-organized Complexity Unfolding Model, there are complex feedforward and feedback relations between the industry configuration and organizations' situations and strategies. An organization's (objective) situation could be very fluid—depending on the relation between the life cycles of the industry and the life cycles of the organization (cf. Christensen's *Innovator's Dilemma*). Therefore, an organization must be judicious in its timing and pace of its interpretation of situation and its formulation of strategies (cf. the earlier discussion of taking time slices).

- Third, any research that aims to describe and understand the self-organized emergence of standards and market configurations must account for the above two points. A little reflection will convince one that the issues of subjective assessment of situations and timing of strategies cannot be analyzed via a general theory. Therefore, the case study is the right method for this line of research.

Organization's Capabilities to Meet Market Needs and Opportunities

As has been pointed out, customers represent one actor group. In fact, it is the desires and needs of the customers that ultimately generate opportunities for creating value, and constitute the market demand that organizations strive to shape and meet. In ICT, technology standardization largely determines the method and the value networks through which customer needs are to be met. Whether a certain technology standard will be adopted depends on whether the products it helps to deliver fit with customers' situations, and whether the value networks it entails fit with the situations of the other actors involved – vendors, complementary service providers, etc. The roles

that an organization has the capacity to play, now and potentially in the future, in (1) providing the technology, (2) influencing standardization, and (3) forming value networks, to meet market needs and opportunities and to create value, is the most fundamental element of an organization's situation.

In organization theory, the capability of an organization encompasses not only the qualities and efficacies of its products and services, but more importantly its ability to perceive and take advantage of market needs and opportunities (Besen and Farrell, 1994; Chiesa and Toletti, 2003; van Wegberg, 2004). In this research model, these aspects of an organization's capacity will be reflected in the organization's interpretation of its situation.

When Sony developed the first videocassette format, Beta, there was a great opportunity to dominate the newly emerging video markets by controlling the technology standard. It is generally agreed that Sony's Beta format was technically superior (e.g. the quality of the recording) to the competing VHS format. However, technology was not the only relevant factor. The Beta format allowed the users to record less than one hour of video, while the users' main desire was to be able to record movies which in general lasted more than one hour (Grindley, 1995). Seen from this light, in spite of the technical superiority of the Beta format, one may say that Sony actually did not have the capability to meet the then emerging market need for video recording (Of course, Sony definitely had the capacity to participate in the VHS value network.).

The extent of an organization's capability objectively circumscribes its strategic options, and its self awareness of its capacity subjectively channels its strategic choice. A firm that lacks the wherewithal to properly perceive or adequately meet market needs and opportunities by itself may make up for its lacuna by playing the role of a follower and joining a standard-setting organization.

The Availability of Complementary Products or Compatibility of Products/Services

ICT products and services are complex and require multiple components and interfaces among them to perform as a system (Tassey, 2000). Products or services that have fewer complementary products are usually less competitive in the market. Moreover, it is virtually impossible to standardize ICT without having complementary products: e.g. computer hardware requires software and vice versa.

In the case of mobile communication networks, handsets are not just complementary products. They need to be compatible with the networks in order to be used. Thus, the available of complementary products or compatibility of product/services should be used depending on the context of industry. The availability of complementary products or compatibility of products/services is one significant element of organizational situation (Rosen, 1994; Chiesa and Toletti, 2003). Because the right mix of complementary or compatible products may or may not already exist, organizations have to worry about how they may be provided, and in what forms and at which times.

This element of an organization's situation can definitely influence its strategy towards standardization. An organization with fewer resources and capabilities to produce complementary or compatible products by itself is more likely to collaborate with others in the formation of value networks and standard-setting. Also, if the technology that the organization pushes to be a standard is compatible to existing complementary goods or systems, then it will be less of a challenge than if complementary or compatible products have to be created. A greater need for complementary or compatible products and lesser capability to manufacture them internally will generally lead an IPR holder to open its IPR to encourage vendors to manufacture the complementary or compatible products.

The Type of Technology Innovation

Broadly speaking, there are two kinds of technology innovation: continuous and disruptive innovation (Christensen, 1997; Seo and Lee, 2007). Continuous technology innovation refers to the evolution of technologies based on existing assets and capabilities. In general, a continuous innovation can be carried out with the existing configuration of value networks. For example, each new version of Microsoft Windows is a continuous innovation, and does not require major realignment of the organizations in the "Microsoft Windows ecology." On the other hand, disruptive technology innovation entails qualitatively different assets and capabilities. In general, it cannot be realized with the existing value networks, and realignment of organizational relationships is necessary. For example, in the migration from GSM (a 2G mobile communication technology) to WCDMA (a 3G mobile communication technology), new and different assets and capabilities such as new frequency bands, IP networks, and multi-mode handsets are required. A 2G actor (e.g. a chipset supplier) does not necessary have the capability to fill the changed role required in 3G (e.g. supplying integrated monolithic IC).

The kind of technology innovation involved can definitely impact an organization's strategy. If the technology in a potential standard is the result of continuous innovation, organizations can use existing assets and capabilities to implement it, but if it is a disruptive innovation, organizations need to work with the new technical architecture of complementary products and participate in new value networks. Moreover, in the case of disruptive innovation, organizations tend to be more willing to open IPR and to collaborate with other actors in building new value networks and setting standards, especially if they do not have enough resources, capabilities and market position to dictate their terms to others. From the perspective of value, to the extent that an organization can better leverage existing resources and capabilities of the existing value networks to create and capture value with technology innovation, the more advantageous position it is in relation to others.

The Position of an Organization in the Market

The market position of an organization refers to its market share (Besen and Farrell, 1994; Chiesa and Toletti, 2003). There are at least two types of market position (Besen and Farrell, 1994). One is an incumbent that holds significant market share and can leverage power through its size; the other is a minor player that has a small or niche market and thus has little market power (Seo and Desouza, 2006). Minor players can be further subdivided into two types: existing and new entrant. Existing minor players are companies that have held small segments of the market for a while; new entrants are organizations that are just entering the market and are starting with no market share.

For example, Microsoft is an incumbent in the PC operating systems and related markets, while Apple is a minor player. On the other hand, through its innovation in digital media, Apple dominates the digital music market, and is currently leveraging this dominance to the video and phone markets. This example illustrates that an incumbent in one market does not necessarily have power in another market unless the two markets share very similar resources and capabilities.

Market position can have a great impact on an organization's standards strategy. An organization that has a strong market position is likely to standardize its technology by itself as a de facto standard because its market position gives it power over others in the technology standardization process (Besen and Farrell, 1994; Chiesa and Toletti, 2003; van Wegberg, 2004). This is especially true when its existing technology is dominant and has locked in adopters. This is the case of Microsoft's position in operating systems for personal computers. This dominant market position allows the organization to control its IPR

over the technology tightly, because it doesn't need to open its technology in order to make it into a standard.

It should be noted, however, that value networks in ICT often span multiple markets. For example, the value networks of mobile communications span the carrier market, the handset market, the Integrated Circuit (IC) market, the electronics manufacturing market, etc. Because technology innovation and standardization shape the value networks, the organizations in all the markets (particularly dominant actors) will try to influence or control the standardization process. Whether they will be successful depends very much on the distribution of power in the value networks, which in turn depends on the next two elements to be discussed—namely, (E) the availability of alternative or substitutable technologies, and (F) the ownership and deployment of intellectual property rights.

The Availability of Alternative or Substitutable Technologies

The term "availability" refers to the possible existence of competing technologies that can provide or support similar functionalities of products or services (Bores et al., 2003). The extent to which substitution is possible depends on the level of technology being considered—the architecture level, the system and core component level, or the supporting component level. Also, the extent to which substitution might be contemplated depends on the time frame—short versus long term. The following example will illustrate these points.

For 2G mobile technology, outside of Europe which mandates GSM, a mobile service carrier must decide between GSM and CDMA, the two dominant technologies which act as alternatives to one another. However, once the carrier has chosen to go with one or the other, they are no longer viable alternatives throughout the whole 2G era because the switching cost is so high. Once the technology architecture has been chosen, say

GSM, there are alternatives to the structuring of the network (using Nokia technology, Siemens', or Ericsson's, etc.), and there are even broader alternatives among handsets. At the time of transition from 2G to 3G and beyond (cf. the notion of time-slice), a carrier again has alternatives. It has the choice of the technology architecture of WCDMA, CDMA2000, WiMax, and to a lesser extent WiFi.

In the future, there will likely be no competing architectures but only a single architecture – *viz.* all mobile communication will be run on an IP (Internet protocol) core through the established VOIP standard, with the technologies of WCDMA, CDMA2000, etc. as alternative and substitutable radio access to the IP core. [Indeed, the ready substitution of radio access is the rationale behind the newer phones that have the dual mode of WiFi and WCDMA.]

In general, the more available the alternative or substitutable technologies are, the more intense the market competition is. This is due to the existence of a greater number of incompatible products in the market (van Wegberg, 2004). However, such intense competition tends to preclude building economies of scale and thus increases the cost of doing business (David and Greenstein, 1990). Therefore, at least at the higher levels of technologies (i.e., core systems/components and architecture), there is a drive towards standardization. But then the existence of alternatives drives intense competition to the venues of (1) setting standards and (2) influencing market adoption of competing standards.

All in all, the availability of alternatives is a very complex element of the situation an organization faces because it spans multiple levels of technology, multiple businesses, and multiple time frames. Moreover, how the alternative or substitutable technologies are made available depends on the ownership and deployment of the actors' IPR.

Ownership and Deployment of IP (Intellectual Property)

One important characteristic of the ICT industry is that its products and services are in general assemblies of functionalities of manifold components and processes unified under certain architectures. These components and processes are in general supplied by different companies, while the architecture can be open or proprietary (owned by the system integrator).

Loosely speaking, the term IP (intellectual property) refers to the "knowledge of how to deliver certain functionality." This knowledge is regarded as property because it is legally recognized and protected by means of copyrights, trade secrets, and patents. In theory, owning IPR (intellectual property rights) means that an organization has certain legal monopoly on delivering functionality through a particular method. But in practice, this monopoly is worthless (1) if nobody is interested in deploying this particular method, or (2) if this monopoly may be easily evaded – by legal means or otherwise, or (3) if the organization somehow fails to monetarize its IPR in the market.

There are two kinds of IP (Intellectual Property) that we need to consider in relation to technology standardization (Mak, 2006). One is essential IP, where the technology is used as a core component in a product or system. The other is non-essential IP, where the technology is for a subsidiary component. Core components are necessary parts in the product or system, while a subsidiary component can be replaced by another component that provides same or similar functionalities. For example, a microprocessor is a core component and a 5.25-inch compact floppy disk drive is a subsidiary component in a PC. The microprocessor is necessary but a 5.25-inch compact floppy disk drive can be replaced by 3-inch compact floppy disk drive, USB drive, or other technologies for removable memory. An organization that holds essential IP is therefore in a more strategically advantageous position than a non-essential IP holder.

Another thing we need to acknowledge about IP is that it is not easily protected (Shell, 2004). Unlike physical property, intellectual property is intangible and it is difficult to prove that someone has infringed on the intellectual property of the original creator. Moreover, the market for IPR is not well established—it is difficult to valuate IP and even more difficult to enforce equitable transactions in IP.

In the ICT industry, every organization is concerned with the potential value of its portfolio of IP *versus* those of other organizations, and how it may leverage its IPR to capture value from the unfolding of the industry's value networks. Thus, correct assessment of one's IPR situation is a critical part of formulating an organization's standards strategy through which it strives to shape a favorable value network for the deployment of its IPs.

The analysis of these fundamental elements makes it possible to understand an organization's situation. Each organization (actor) interprets its own situation based on its orientation and thus decides on its strategies. The next section introduces a perspective in organizational interpretation of situation.

3.2. Organizational Perspective for Interpreting the Situation

Although the six situational elements can be objectively evaluated by outsiders, these elements are subjective matters to the organization in the situation, similar to how each individual perceives his/her surroundings subjectively while an observer of this individual sees her/his surroundings objectively. However, when the observer tries to comprehend the individual's behaviors, the observer needs to consider the logic and culture of the individual to analyze how (s)he interprets the situation subjectively. This would lead us to

infer that the observer needs to understand the individual's context.

Like individuals, actors (organizations) interpret their situations differently based on the possibility of value that they can create or capture, because of the subjective and context-specific nature of their perception of value (Lepak et al., 2007). Actors interpret their situations differently based on their logic and culture, which have formed over time, as human beings tend to do (Anderson, 1999). This implies that actors in similar situations can interpret their situation very differently and make different decisions about their strategies.

This perspective is supported by the view of Organization Learning as well. An organization as an adaptive entity senses changes in its surrounding and adapts accordingly (Argote, 1999; Argyris, and Schon, 1996). Thus, it develops its own logic and culture to sense and respond to its changing situation over time. Logic refers to what and how things should be done. Culture implies commonly shared values and virtue. Based on its logic and culture, an organization's orientation to interpret its changing situation is formed in certain ways as it senses and adapts to the dynamic environment. In this view, not only is it possible for actors in similar situations to interpret their situations differently, but also for the same actor to interpret similar situations differently over time, because logic, culture and orientation evolve.

One example illustrates the importance and role of organizational interpretation. U.S. manufacturers such as Intel, Micron Technology, and Texas Instruments enjoyed great value from the cutting edge technology of the computer memory chip in the 1960s and 1970s (Shell, 2004). The exchange value decreased when more companies introduced similar products, especially when cheap Korean and Japanese chips flooding into the U.S. in the 1980s (Shell, 2004). Consequently, the American companies suffered painful losses. Each of the U.S. companies found themselves in a similar situation, but their interpretations and therefore their strategies were drastically different. Intel, with its strong capabilities in designing and manufacturing chips, interpreted that there was less value in the memory-chip market because of high competition from foreign companies, and therefore decided to exit this market altogether and focus only on microprocessors (Shell, 2004). However, Micron Technology, which had competence in the political and legal realms, interpreted the situation differently and decided to advocate for political and legal barriers against foreign companies. It filed antidumping complaints against the Korean and Japanese companies, so it could sustain its U.S. market share (Shell, 2004). Texas Instruments had its own interpretation as well. Because it owned the earliest patents as a pioneer in the chip technology and recognized the value of its IPR, it decided to use IPR as the basis to attack the foreign companies (Shell, 2004). With financial resources collected by license agreements, Texas Instruments was able to develop another competence, namely the design and manufacture of digital signal processors for other potential markets.

Although it is impossible to fully understand an organization's interpretation of its situation as an outside investigator because of the difficulties of analyzing an organization's logic, culture, and orientation, the perspective of value creation and capture provides a reasonable approach. At the very least, we can reasonably say that an actor must evaluate the situation with the goal of creating and capturing value: (1) what value its capabilities and resources add or create for existing products or services; (2) what situational elements are related to current value and whether value can be sustained in potential future markets; (3) whether its situation is advantageous for creating or capturing value for possible market needs and opportunities; (4) what resources or capabilities it needs for value creation and how it can gain them. As shown in the example, it is impossible to discern the underlying logic, culture, and orientation of Intel, Micron Technology and Texas Instruments. However, it is possible to analyze organizational interpretation by identifying each organization's

approach to value creation and capture. Intel's interpretation of the situation was that the value of memory chips decreased due to competition from foreign companies, and that greater value could be more usefully pursued through microprocessors. Micron Technology thought that the value of memory chips was worth defending by building barriers through close interactions with the U.S. government. Texas Instruments took advantage of the situation to add to the value of memory chips by charging license fees to the foreign companies, so that it could develop its competence for other types of chips for other markets.

As an outside investigator, we can evaluate an actor's interpretation of the situation from the perspective of value creation and capture through the decisions and behaviors that the actor makes. Technology standardization can offer great opportunities for an actor to create or capture value through interpretation of its situation regarding the control of standards. For example, if value rests in a core component that an organization owns, it will be better to standardize the architecture of technology as an open standard so more actors can adopt it quickly, while still maintaining value by owning the core component. This was the case for the PC OS for Microsoft and microprocessors for Intel. Therefore, organizational situation and the way an actor interprets its situation affect organization standards strategy directly.

3.3. Aspects of Standards Strategies

As pointed out in the section on Self-organized Complexity and ANT, it is necessary to investigate an actor's context to analyze its strategies and behaviors. Analyzing the organizational situation and the organization's interpretation of its situation is one way to understand organizational context. Organizations form and shape their standards strategies within the context of their situations and interpretations.

The ICT standards that organizations pursue affect how value networks form and evolve. For organizations, building a strong value network is the foundation for a competitive market position to deliver services or products. Thus, the formation of the value network is an important aspect in organizations' strategies.

Because of the existence of political regions and the compatibility requirements of communications technologies, standard-setting organizations have been used in some places such as Western Europe. Using the standard-setting organizations is one way to form and shape the value network. Thus, issues related to standard-setting organizations are strategic decisions for organizations.

Since IPR became significant in business, the value of organizations' Intellectual Properties (IP) has increased. Organizations strategically build, manage, and utilize their IP to attract other actors to their networks and suppress the business activities of competitors. Thus, organizational strategy in IPR plays a significant role in technology standards.

In this section, these three aspects of standards strategy that emerge from the literature on standards will be presented in depth.

Configuration of Value Network

Is important in determining an organization's ability to deliver complete products or services to customers. The concept of the value network is rooted in the supply chain, which refers to the sequence or processes for organizations to create products and services, from procuring raw materials to manufacturing in one or more factories, to storing them in one or more warehouses, to shipping to retailers or customers (Simchi-Levi et al., 2003). Supply chain management has always been an important issue in business, because the efficient integration of suppliers, manufacturers, distributors, and retailers affects the cost of products and the conformation to customer requirements (Simchi-Levi et al., 2003).

Researchers and practitioners have re-evaluated the concept of the supply chain as a way to create value and started to use the term "value chain" to emphasize each part of the supply chain as value-adding activities (e.g. Porter, 1985). The reason that value "network" is used instead of "chain" in this book is that "chain" implies a linear flow of value-adding activity, while "network" can include actors not directly on the supply chain whose activities indirectly add value as well, such as complementary products (Verna, 2002). "Chain" is also used for rather simple industries and focuses on one or a few companies, such as Walmart's supply chain, while "network" is more appropriate for complex industries such as the information and communications technology industry that includes many organizations. For example, when the VHS videocassette format was competing to be the standard, the greater availability of content in this format, such as movies released by movie-distributors, added great value to VHS players, even though movie-distributors did not directly participate in the value chain of VHS player production. Thus, reflecting the complexity of the standardization phenomenon, the term "value network" is more appropriate. Adopting the broader concept of Verna (2002), the definition of value network used here is as follows:

Value network is any Web of relationships that generates value directly or indirectly through complex dynamic exchanges between actors (p. 6).

In this sense, end-customers also generate value in terms of identifying and informing value directly or indirectly to other actors. For organizations, their successes rest on the proper appreciation of use and exchange value for customers, and the strength of their capabilities to deliver goods to meet customers' needs through forming efficient and effective value networks (Kang et al., 2007; Priem, 2007).

The configuration of value networks is the formation of manifold networks among actors in the proposed model. There are competing and complementary networks. The successful value network configuration will be the strong network that emerges from interactions among actors that adds value from all participants and brings benefits to the participants as a reward. Thus, it is a fundamentally significant aspect of strategies for actors.

The configuration of value networks requires not only technological configurations but also business agreements between companies to provide components, interfaces and complementary products. In an organization's situation, the technology innovativeness, the availability of complementary products or compatibility of products/services, and organizational market position directly impact an organization's standardization strategies as reflected in potential configurations of value network. The organization with great resources, including capabilities, that can use an existing value network is in a more advantageous position than organizations that need to form a new value network and have less power. To standardize a technology and sustain it in the post-standardization period, the organization needs strategy to build and maintain a strong value network that can compete against others.

Formation of Standard-Setting

Refers to whether and how an organization will collaborate with others to standardize a technology. As shown in the literature review, there are various types of standard-setting processes: voluntary collaboration between companies like strategic alliances, informal or formal standard-setting organizations sponsored by one or more entities, and stand-alone standard-setting without any collaboration.

This aspect of organizational standards strategy is also about shaping a value network through interactions between actors that influence indus-trial configurations or patterns in the complex phenomenon of standardization. An organization should utilize appropriate standard-setting collaborations to create or capture existing and potential value. If an organization has a strong market position and an effective value network for complementary products, and can use those existing resources and capabilities to create value by standardizing its technology, it does not need help from other actors. Then it might pursue standardization by itself.

Openness of IPR (Intellectual Property Rights)

Refers to how much the IPR of a technology is opened to other organizations. This aspect is not a simple matter of open or closed; rather there are various degrees and types of openness. IPR can be opened totally without any license fees, or it can be partially opened in several different ways: open licensing is the practice of giving a license to every actor for the same fee; cross-licensing is an agreement between two actors to share the IPR of technologies that each actor holds; patent pooling is sharing licenses among a group of IPR holders whose technologies are needed for one complete product or system and splitting license fees that are charged to others as the pooling fees of all IPR (Shell, 2004). Then there is the possibility of not opening IPR at all, and also the option of defending the technology as a secret without even patenting it.

These various ways to trade IPR become sources of value creation and capture for actors in shaping value or standard-setting networks. Thus, organizational IPR strategy affects interactions between actors in the formation of networks in the model. As Blind and Thumm (2004) and Lea and Hall (2004) point out in their studies, strong IPR holders are more hesitant to join a standard-setting organization due to the fear of sharing their IPR with other members. It also affects value network configuration and the bandwagon effect.

For example, JVC expedited the configuration of its value network and stimulated the bandwagon effect by licensing VHS technology to other actors in the standardization of videocassette format, while Sony followed the strategy of keeping its Beta technology closed (Grindley, 1995).

Depending on its situation and its interpretation of the situation, an organization needs to formulate its IPR strategy. If it does not have strong market position, sufficient resources and capabilities to produce complementary products, or the knowledge to analyze market opportunities and needs, it is more likely to open up its IPR to invite potential collaborators.

This aspect may be less important for organizations in the position of technology adopters rather than critical IPR holders, but they still need to consider how to access IPR before they commit to adopting one technology as a standard. Once an actor adopts one technology over another, its interaction with the IPR holder becomes a part of the network to standardize the same technology. This commitment can be critical for the adopter, especially when the switching cost to another technology cannot be recuperated.

The value aspect of IPR strategies is very important for analyzing organizational standards strategy not only during the standardization process but also in the post-standardization phase. An actor can leverage IPR in the configuration and shaping of various networks to create or add value during the standardization process. In the post-standardization phase, if IPR is not totally opened, the value created through standardization can be captured for a certain period of time instead of immediately being worn away. The introduction of similar products from competitors may decrease the value created through standardization, but IPR can be used to isolate the value by preventing the product from being copied. Capturing value through this process is called an isolating mechanism (Lepak et al., 2007). IPR is a great tool to capture value through the isolating mechanism in the post-standardization phase.

3.4. Summary of Framework to Analyze Standards Strategy

ICT standardization is a phenomenon of self-organized complexity. To comprehend and analyze this complexity, the general model, Self-organized Complexity Unfolding Model is proposed. The complexity comes out of many actors' involvement and networks that are formed by the interactions between actors. The industrial configuration or pattern is the result of the aggregated interactions of networks that can be newly formed, evolved from the existing configuration, or destroyed by other networks. Each actor cannot by itself shape the industry-wide configuration to its competitive advantage, but it can possibly affect the ongoing activity of value creation and capture through value networks, because the aggregated networks eventually become the industrial configuration or pattern. Here is where organizational standards strategy plays a significant role in forming and shaping networks, because the actors' networks are vehicles for organizations to execute their strategies.

To analyze organizational standards strategy, we need to understand the context of actors. Each actor has a dynamic situation. The actor interprets its situation subjectively and tries to fit itself into value network through creating and capturing value. However, value creation and capture for ICT standards do not depend on just one single actor, but on coordination between and agreements among actors. Thus, actors try to form and shape standards strategies within their contexts that can create a favorable future value network for themselves. To create this potential network, actors utilize standard-setting organizations and maneuvers with intellectual property rights.

As mentioned, ICT standards have had significant socio-economic impact in ICT-related markets; consequently, organizational standards strategy has become a major part of business strategy in ICT.

To analyze actors' standards strategies, the Framework of Organizational Standards Strategy is suggested. In the Framework, six elements are used to analyze organizational situation and the value-based view is applied to understand an organization's interpretation of its situation. Based on the situation and interpretation, organizational strategies are analyzed from three aspects. These three aspects of standards strategy are not independent of each other, but related. Strong IPR can control a value network but also can reduce the availability of complementary products or compatibility of products/services, creating a less compatible product that cannot bring forth the necessary economic factors for standardization. The formation of standard-setting can influence the configuration of value network and vice versa, because those interactions between actors are meant to shape strong networks that win over other networks in the overall picture of self-organized complexity.

Using this general model and framework, we can analyze why a particular organizational standards strategy is formed and how it affects the formation of self-organization in the complex ICT standardization phenomenon.

REFERENCES

Amabile, T. M. (1996). *Creativity in context: Update to the social psychology of creativity*. Boulder, CO: Westview Press.

Anderson, P. (1999). Complexity theory and organization science. *Organization Science, 10*(3), 216–232. doi:10.1287/orsc.10.3.216.

Argote, L. (1999). *Organizational learning: Creating, retaining, and transferring knowledge*. Boston, MA: Kluwer Academic.

Argyris, C., & Schon, D. A. (1996). *Organizational learning II*. Boston, MA: Addison Wesley.

Ashby, W. R. (1947). Principles of the self-organizing dynamic system. *The Journal of General Psychology, 37*, 125–128. doi:10.1080/00221309.1947.9918144 PMID:20270223.

Barney, J. B. (1991). Firm resources and sustained competitive advantage. *Journal of Management, 17*(1), 99–120. doi:10.1177/014920639101700108.

Benbya, H., & McKelvey, B. (2006). Using co-evolutionary and complexity theories to improve IS alignment: A multi-level approach. *Journal of Information Technology, 21*(4), 284–298. doi:10.1057/palgrave.jit.2000080.

Besen, S. M., & Farrell, J. (1994). Choosing how to compete: Strategies and tactics in standardization. *The Journal of Economic Perspectives, 8*(2), 117–131. doi:10.1257/jep.8.2.117.

Blackman, C. R. (1998). Convergence between telecommunications and other media: How should regulation adapt? *Telecommunications Policy, 22*(3), 163–170. doi:10.1016/S0308-5961(98)00003-2.

Blind, K., & Thumm, N. (2004). Interrelation between patenting and standardisation strategies: Empirical evidence and policy implications. *Research Policy, 33*(10), 1583–1598. doi:10.1016/j.respol.2004.08.007.

Bores, C., Saurina, C., & Torres, R. (2003). Technological convergence: A strategic perspective. *Technovation, 23*(1), 1–13. doi:10.1016/S0166-4972(01)00094-3.

Bowman, C., & Ambrosini, V. (2000). Value creation versus value capture: Towards a coherent definition of value in strategy. *British Journal of Management, 11*(1), 1–15. doi:10.1111/1467-8551.00147.

Callon, M. (1986). Some elements of a sociology of translation: Domestication of the scallops and the fishermen. In Law, J. (Ed.), *Power, Action and Belief: A New Sociology of Knowledge?* London: Routledge & Kegan Paul.

Camazine, S., Deneubourg, J., Franks, N. R., Sneyd, J., Theraulaz, G., & Bonabeau, E. (Eds.). (2003). *Self-organization in biological systems*. Princeton, NJ: Princeton University Press.

Chiesa, V., & Toletti, G. (2003). Standard-setting strategies in the multimedia sector. *International Journal of Innovation Management, 7*(3), 281–308. doi:10.1142/S1363919603000829.

Christensen, C. M. (1997). *The innovator's dilemma: When new technologies cause great firms to fail*. Boston, MA: Harvard Business School Press.

Cunha, E, M. P., & Da Cunha, J. V. (2006). Towards a complexity theory of strategy. *Management Decision, 44*(7), 839–850. doi:10.1108/00251740610680550.

David, P. A., & Greenstein, S. (1990). The economics of compatibility standards: An introduction to recent research. *Economics of Innovation and New Technology, 1*, 3–41. doi:10.1080/10438599000000002.

David, P. A., & Steinmueller, W. E. (1994). Economics of compatibility standards and competition in telecommunication networks. *Information Economics and Policy, 6*(3-4), 217–241. doi:10.1016/0167-6245(94)90003-5.

Drazin, R., & Sandelands, L. (1992). Autogenesis: A perspective on the process of organizing. *Organization Science, 3*(2), 230–249. doi:10.1287/orsc.3.2.230.

Estep, M. (2006). *Self-organizing natural intelligence: Issues of knowing, meaning, and complexity*. Dordrecht, The Netherlands: Springer.

Gao, P. (2005). Using actor-network theory to analyse strategy formulation. *Information Systems Journal, 15*(3), 255–275. doi:10.1111/j.1365-2575.2005.00197.x.

Goldbaum, D. (2006). Self-organization and the persistence of noise in financial markets. *Journal of Economic Dynamics & Control, 30*(9/10), 1837–1855. doi:10.1016/j.jedc.2005.08.015.

Graham, I., Spinardi, G., Williams, R., & Webster, J. (1995). The dynamics of EDI standards development. *Technology Analysis and Strategic Management, 7*(1), 3–20. doi:10.1080/09537329508524192.

Grant, R. M. (1991). The resource-based theory of competitive advantage: Implications for strategy formulation. *California Management Review, 33*(1), 114–135. doi:10.2307/41166664.

Grindley, P. (1995). *Standards, strategy, and policy: Cases and stories*. Oxford, UK: The University Press. doi:10.1093/acprof:oso/9780198288077.001.0001.

Haken, H. (2006). *Information and self-organization: A macroscopic approach to complex systems* (3rd ed.). New York: Springer.

Hanseth, O., Jacucci, E., Grisot, M., & Aanestad, M. (2006). Reflexive standardization: Side effects and complexity in standard making. *Management Information Systems Quarterly, 30*, 563–581.

Kaghan, W. N., & Bowker, G. C. (2001). Out of machine age? Complexity, sociotechnical systems and actor network theory. *Journal of Engineering and Technology Management, 18*(3/4), 253–269. doi:10.1016/S0923-4748(01)00037-6.

Kang, S. C., Morris, S. S., & Snell, S. A. (2007). Relational archetypes, organizational learning, and value creation: Extending the human resource architecture. *Academy of Management Review, 32*(1), 236–256. doi:10.5465/AMR.2007.23464060.

Katz, M. L., & Shapiro, C. (1985). Network externalities, competition, and compatibility. *The American Economic Review, 75*(3), 424–440.

Krugman, P. R. (1996). *The self-organizing economy*. Cambridge, MA: Blackwell Publishers.

Latour, B. (1987). *Science in action: How to follow scientist and engineers through society*. Cambridge, MA: Harvard University Press.

Latour, B. (1996). *Aramis, or the love of technology*. Cambridge, MA: Harvard University Press.

Lea, G., & Hall, P. (2004). Standards and intellectual property rights: An economic and legal perspective. *Information Economics and Policy, 16*(1), 67–89. doi:10.1016/j.infoecopol.2003.09.005.

Lea, M., O'Shea, T., & Fung, P. (1995). Constructing the networked organization: Content and context in the development of electronic communications. *Organization Science, 6*(4), 462–478. doi:10.1287/orsc.6.4.462.

Lepak, D. P., Smith, K. G., & Taylor, M. S. (2007). Value creation and value capture: A multilevel perspective. *Academy of Management Review, 32*(1), 180–194. doi:10.5465/AMR.2007.23464011.

Liebowitz, S. J., & Margolis, S. E. (1994). Network externality: An uncommon tragedy. *The Journal of Economic Perspectives, 8*(2), 133–150. doi:10.1257/jep.8.2.133.

Mak, K. T. (2006). *Standard strategy for high-tech business*. Chicago, IL: University of Illinois at Chicago.

Markus, M. L., Steinfield, C. W., Wigand, R. T., & Minton, G. (2006). Industry-wide information systems standardization as collective action: The case of the U.S. residential mortgage industry. *Management Information Systems Quarterly, 30*, 439–465.

Morel, B., & Ramanujam, R. (1999). Through the looking glass of complexity: The dynamics of organizations as adaptive and evolving systems. *Organization Science, 10*(3), 278–293. doi:10.1287/orsc.10.3.278.

Penrose, E. T. (1959). *The theory of the growth of the firm*. New York: Wiley.

Porter, M. E. (1980). *Competitive strategy*. New York: The Free Press.

Porter, M. E. (1985). *Competitive advantage: Creating and sustaining superior performance*. New York: The Free Press.

Post, J. E., Preston, L. E., & Sachs, S. (2002). *Redefining the corporation: Stakeholder management and organizational wealth. Stanford, CA*. Stanford: Business Books.

Priem, R. L. (2007). A consumer perspective on value creation. *Academy of Management Review, 32*(1), 219–235. doi:10.5465/AMR.2007.23464055.

Rohracher, H. (2003). The role of users in the social shaping of environmental technologies, innovation. *European Journal of Soil Science, 16*(2), 177–192.

Rosen, B. N. (1994). The standard setter's dilemma: Standards and strategies for new technology in a dynamic environment. *Industrial Marketing Management, 23*(3), 181–190. doi:10.1016/0019-8501(94)90031-0.

Sarker, S., Sarker, S., & Sidorova, A. (2006). Understanding business process change failure: An actor-network perspective. *Journal of Management Information Systems, 23*(1), 51–86. doi:10.2753/MIS0742-1222230102.

Seo, D., & Desouza, K. (2006). Power-shifting. *Business Strategy Review, 17*(1), 26–31. doi:10.1111/j.0955-6419.2006.00387.x.

Seo, D., & Lee, J. (2007). Gaining competitive advantage through value-shifts: A case of the South Korean wireless communications industry. *International Journal of Information Management, 27*(1), 49–56. doi:10.1016/j.ijinfomgt.2006.12.002.

Seo, D., & Mak, K. T. (2010). Using the thread-fabric perspective to analyze industry dynamics: An exploratory investigation of the wireless telecommunications industry. *Communications of the ACM, 53*(1), 121–125. doi:10.1145/1629175.1629205.

Shell, G. R. (2004). *Make the rules or your rivals will*. New York: Crown Business.

Simchi-Levi, D., Kaminsky, P., & Simchi-Levi, E. (2003). *Designing and managing the supply chain: Concepts, strategies, and case studies*. New York: McGraw-Hill.

Simon, H. A. (1996). *The sciences of the artificial*. Cambridge, MA: MIT Press.

Sirmon, D. G., Hitt, M. A., & Ireland, R. D. (2007). Managing firm resources in dynamic environments to create value: Looking inside the black box. *Academy of Management Review, 32*(1), 273–292. doi:10.5465/AMR.2007.23466005.

Slater, S. F. (1997). Developing a customer value-based theory of the firm. *Journal of the Academy of Marketing Science, 25*(2), 162–167. doi:10.1007/BF02894352.

Smith, A. C. T., & Graetz, F. (2006). Complexity theory and organizing form dualities. *Management Decision, 44*(7), 851–870. doi:10.1108/00251740610680569.

Stanley, H. E., Amaral, L. A. N., Buldyrev, S. V., Gopikrishnan, P., Plerou, V., & Salinger, M. A. (2002). Self-organized complexity in economics and finance. *Proceeding of the National Academy of Sciences of the United States of America, 99*(1), 2561-2565.

Tassey, G. (2000). Standardization in technology-based markets. *Research Policy, 29*(4-5), 587–602. doi:10.1016/S0048-7333(99)00091-8.

Thompson, J. D. (1967). *Organizations in action: Social science bases of administrative theory*. New York: McGraw-Hill.

Turcotte, D. L., Malamud, B. D., Guzzetti, F., & Reichenbach, P. (2002). Self-organization, the cascade model, and natural hazards. *Proceeding of the National Academy of Sciences of the United States of America, 99*(1), 2530-2537.

Turcotte, D. L., & Rundle, J. B. (2002). Self-organized Complexity in the physical, biological, and social sciences. *Proceeding of the National Academy of Sciences of the United States of America, 99*(1), 2463-2465.

van Wegberg, M. (2004). Standardization process of systems technologies: Creating a balance between competition and cooperation. *Technology Analysis and Strategic Management, 16*(4), 457–478. doi:10.1080/0953732042000295784.

Verna, A. (2002). *A value network approach for modeling and measuring intangibles*. Retrieved from http://www.vernaallee.com/value_networks/A_ValueNetwork_Approach.pdf

Von Neumann, J., & Morgenstern, O. (1953). *Theory of games and economic behavior* (3rd ed.). Princeton, NJ: Princeton University Press.

Walsham, G., & Sahay, S. (1999). GIS for district-level administration in India: Problems and opportunities. *Management Information Systems Quarterly, 23*(1), 39–65. doi:10.2307/249409.

Wernerfelt, B. (1984). A resource-based view of the firm. *Strategic Management Journal, 5*(2), 171–180. doi:10.1002/smj.4250050207.

Woodruff, R. B. (1997). Customer value: The next source for competitive advantage. *Journal of the Academy of Marketing Science, 25*(2), 139–153. doi:10.1007/BF02894350.

Chapter 3
Preface to the Research on Standards in the Mobile Communications Industry

EXECUTIVE SUMMARY

In this chapter, the Self-Organized Complexity Unfolding Model and the Framework of Organizational Standards Strategy are applied and explored further to the evolution of standards in the mobile communications industry. To introduce the mobile communications industry, three different infrastructures that presently exist at the beginning of the 21st century for communication over long distances are discussed. The reason to present them is to illustrate the position of the mobile communications industry in our society and how it is related to other industries.

1. THREE DIFFERENT INFRASTRUCTURES

First is the landline infrastructure, which has the longest history. This infrastructure includes telephone service providers that implement and maintain the physical network, systems manufacturers that produce systems for telephone service providers, and users.

Second is the mobile communications infrastructure. Mobile communications infrastructure requires more sophisticated techniques than landline infrastructure to provide mobility. Service

providers and manufacturers for the mobile communication infrastructure are similar to those for landline infrastructure. Although some providers have both landline and mobile communications infrastructures and some manufacturers produce systems for both infrastructures, many new actors have entered the market since the dawn of the mobile communications industry. Another thing to notice is that the landline and mobile communications infrastructures are interrelated in order to facilitate calls between mobile handsets and landline telephones. More detailed information about mobile technological systems will be provided in the section of Technological Background in Chapter 4.

DOI: 10.4018/978-1-4666-4074-0.ch003

Third is the Internet network based on different media (e.g., Digital Subscriber Loop [DSL], fiber optic cable, etc.). Many actors from the computing industry, which are typically distinct from those in the communications industry, are involved in establishing the Internet network. With Voice over IP, many Internet service providers are able to provide services similar to what communications service providers have traditionally provided. Manufacturers for the Internet network systems (e.g. Intel and Cisco) have different backgrounds from those in the communications industry. This infrastructure has begun to interact with landline and mobile communications infrastructure in the converging network.

2. EVOLUTION OF MARKET

Before investigating organizational standards strategies that have affected the evolution of self-organized configuration in the mobile communications industry, it is worth understanding the evolution of the market at the industrial level. In the developed countries, governments and companies first concentrated on building national landline networks before the advent of mobile communications technology. In this nascent, pre-1G (first generation) technology stage, mobile communications services were very limited, with mostly manual systems. The 1G period opened up the mobile communications market for commercial use by providing mobility to users, but the service was still expensive and the capacity of the systems was limited. Thus, the 1G market was restricted to a very small number of customers. The evolution of the market from 1G to 2G created a new mass market with affordable prices for all kinds of mobile users from business people to young students. In the developed countries, competition among mobile service providers became intense as 2G matured, bringing prices down. The mobile service providers started to seek new sources of revenue, and there was a new market need for data delivery when the Internet market expanded rapidly

in the late 1990s and early 2000s. Thus, the market is now evolving toward 3G, offering different kinds of services other than voice delivery, so service providers can increase or at least maintain ARPU (Average Revenue Per User).

3. EVOLUTION OF TECHNOLOGY

The market has leapt forward at each stage with each new generation of technology. Without technological supports, it is impossible for the market to evolve as it has. Although it is arguable whether markets have driven technologies or vice versa, there is no doubt that there are mutual influences between technology and market. The 1G mobile communications technologies such as NMT and AMPS (Advanced Mobile Phone System) were based on analogy and cellular system, which was more sophisticated than the primitive technologies used before the 1G period. However, 1G technologies still had technical limits. The most distinctive difference between 1G and 2G was that 2G mobile communications technologies (e.g. GSM and IS-95) were digital, which increased capacity and quality and required lower battery power for handsets. 2G technologies provided services and products with lower prices for a vastly increased number of users. In contrast, 3G mobile communications technologies (e.g. WCDMA and CDMA2000) have focused on increasing bandwidth capacity to provide more advanced services, for example, Internet browsing.

The development and implementation of a mobile communications technology requires greater resources and capabilities than any single organization can afford, because all systems (e.g. mobile devices, base station systems and switching systems) have to be compatible with each other to provide a seamless service. For this reasons, actors in the industry need to cooperate and standardize a mobile communications technology, so they can provide services and products at affordable prices to create a mass market.

4. EVOLUTION OF MOBILE COMMUNICATIONS TECHNOLOGY STANDARDS

Organizations may gain competitive advantage by participating in and influencing standard-setting processes and shaping value networks. In addition, the actors that are able to standardize their technology in a large market may obtain a position to leverage their installed-base market not only in the existing generation but also for the next generation. Thus, the organizations that can influence the industrial evolution through their strategies in standard-setting processes and shaping value networks are the major actors.

To analyze organizational standards strategies in the evolution of mobile communications, we should consider the evolutions of market and technology innovation along with the roles of technology standards and the changes of value networks. They are very complex phenomena to comprehend. Thus, the proposed Self-organized Complexity Unfolding Model and the Framework of Organizational Standards Strategy for ICT will be validated in their viability through their application to these complicated and dynamic phenomena.

The goals of the following case studies are: (1) to apply the proposed model and framework to this continuously evolving industry to illustrate their applicability and viability; (2) to refine the proposed model and framework if necessary; (3) to introduce a longitudinal two-level (industrial and organizational levels) approach to analyze organizational standards strategy instead of the cross-sectional single-level analysis that most researchers have adopted; and (4) to provide fundamental knowledge about the evolution of mobile communications technology standards and the strategies of the actors involved.

At the industrial level (see Figure 1), there was the evolution of industrial self-organized configuration that influenced the situation of each actor. The evolution was also affected by the strategies and behaviors of major actors. Thus, Self-organized Complexity Unfolding Model will be applied to analyze the industrial level and the feedback and feed-forward relationships with the

Figure 1. The evolution of mobile communications technology standards

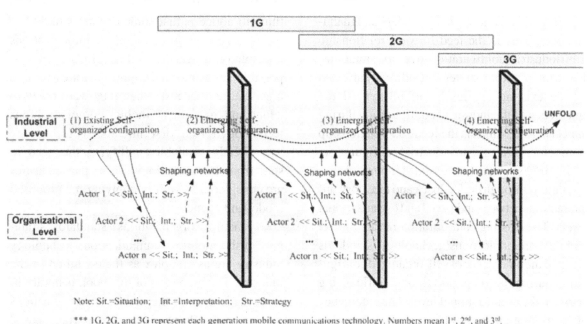

45

organizational level. The feed-forward implies that the industrial configuration influences each actor's situation. Actors shape their strategies based on their own situations and interpretations. These strategies interact with each other and affect the evolution of the industrial configuration, which leads to feedback.

For example, Self-organized Complexity Unfolding Model will be used to explain: (1) how the mobile communications industry evolved from markets segmented by nation before the first generation (1G) mobile communications technology to still segmented, but at least united among Nordic countries, markets during the 1G period; (2) how the less segmented 1G mobile communications industry continuously evolved to ultimately two separate markets using GSM and IS-95 (CDMA) technology standards during the second generation (2G) mobile communications technology period; and (3) how the 2G industry evolved into the 3G mobile communications technology standards. The existing industrial configuration before the 1G arena, which was segmented and dominated by a government-owned monopoly organization or few companies like AT&T in the U.S., affected the situations of many actors in the mobile communications industry including dominant service providers and manufacturers. This is the feed-forward relationship from industrial configuration to actors' situations. These major actors' strategies (indirectly affected by the existing configuration) influenced the emergence of the industrial configuration during the 1G period, which is the feedback relationship from actors' strategies to industrial configuration.

Figure 1 shows the evolution of industrial configurations through the feedback and feed-forward interactions between industrial and organizational levels. The existing configuration before the 1G mobile communications technology standards affected the emergence of self-organized configuration during the 1G period through influencing actors at the organizational level. The emergence of configuration in the 1G period became the

existing configuration for the next emerging self-organized configuration.

At the organizational level, how organizations shape their strategies for the upcoming self-organized configuration based on their situations influenced by the existing configuration will be analyzed by Framework of Organizational Standards Strategy. The organizational standards strategies affect the formation of networks. These networks interact with each other, which affect the evolution of the self-organized configuration at the industrial level. From the above example, Framework of Organizational Standards Strategy will be used to answer the following questions – what the situations of major actors (e.g. government-owned mobile service providers in Europe and AT&T in U.S.) were before the 1G period; how they interpreted their situations; and how they arrived at strategies.

Figure 1 provides the conceptual evolution of the mobile communications industry including interactions with organizational situations, interpretation and strategies. The time divisions between 1G and 2G or 2G and 3G partially overlapped, as seen in Figure 1. For example, in the transition period between two generations of technologies, these technologies co-existed for a while. Another thing to notice is that some countries moved to the next generation of mobile communications technology faster than others. Indeed, it was possible for some developing counties to skip a generation, because when they were ready to adopt and implement mobile communications technology for the first time, the 2G technology was already available, nullifying the need to start from 1G technology. These particularities are considered as well in applying the proposed model and framework.

By nature, organizations' standards strategies in the mobile communications technology industry are as complex as the evolution of the industry. The emphasis of this book is not to inform the history of the mobile communications industry, but rather to focus on important actors' standards strategies and the relationship between these strategies and the evolution of the industrial configuration by applying the proposed model and framework. This is not a confirmatory study to prove the proposed model and framework, but an exploratory application to validate the viability of the proposed model and framework. Thus, the proposed model and framework can be improved by applying them to other industries.

With this in mind, major actors will be identified and the proposed model and framework will be applied to analyze: (1) how the existing industrial configuration affected the major actors' situations; (2) how they interpreted the situation to formulate standards strategies; (3) what their strategies were; and (4) how these strategies interacted with each other to influence the emerging industrial configuration.

The case studies will be presented as follows: (1) the background of the mobile communications industry will be explained including historical and technological conditions and major actor groups as a foundation from which to approach the cases; (2) the industrial configuration and a deep analysis of major actors' situations, interpretations and strategies and the interactions between the configuration and the strategies in standardizing each generation (1G through 3G) of mobile communications technology will be presented over three chapters; (3) how the industry has been unfolded beyond the 3G mobile communications technology standards will be explored; and (4) finally, this book will be concluded with findings and limitations.

Chapter 4
Background of the Mobile Communications Industry

EXECUTIVE SUMMARY

Before we dive into how the mobile communications industry has developed, it is important to understand the general background of this industry. In this chapter, the background of the mobile communications industry is explained including historical and technological conditions and major actor groups as a foundation to approach the cases that are presented in chapter 5, 6, 7, and 8.

1. HISTORICAL BACKGROUND

Before the fifteenth century, people exchanged messages between distant regions through various methods such as beacon fires, messengers, and flags. The main purpose of most messages was related to military and sovereign matters of rulers. The development of the mobile communications industry is an extension of the evolution of postal service, telegraphy, and telephony. They were all designed to communicate more quickly over long distances as the socio-economic conditions of human activities changed.

Although individuals invented telephony technologies and founded private telephony companies, governments in many countries nationalized landline telephone networks in the

DOI: 10.4018/978-1-4666-4074-0.ch004

late 1890s and early 1900s through World Wars I and II (Noam, 1992). After Alexander Graham Bell was awarded the U.S. patent for the invention of the telephone in 1876, he and his colleagues found Bell Telephone, which became the American Bell Telephone Company (AT&T) in 1880. AT&T enjoyed great market share as a virtual monopoly in the U.S. telephone industry, even though some small companies shared the market in some regions, until AT&T was broken up in 1984 by anti-trust litigation. Afterwards, the U.S. telephone industry was run by private companies. In contrast, the governments in many other countries ran their national telephone industry through their Post, Telegraphy, and Telephony (PTT) bureaus after nationalizing private companies. This was the situation before the liberalization of mobile communications operators in the 1980s and 1990s. The timing and development of the liberalization varied by nation.

The European Commission started to encourage the liberalization of the mobile device market first in 1988 (Bekkers, 2001), while developing the pan-European 2G mobile communications technology standard. It urged mobile communications system manufacturers to abolish their special and exclusive relationships with mobile service providers. These relationships will be explained in more detail in the section of Background of the 1G Technology Standards in Chapter 5. Consequently, the service providers and customers had more options for mobile communications systems. This made the incumbent manufacturers confront more competition, while it was an opportunity for new entrants to join the mobile communications market. The Commission urged its members to open up the mobile communications service market in 1990 by allowing more mobile service providers into the markets, which had previously been dominated by monopoly PTTs. The liberalization in Europe affected the liberalization of mobile communications industries in other nations.

This historical background, which includes the history of policy changes, has influenced the standardizations of mobile communications technologies. Understanding this background will help to analyze actors' situations, interpretations and strategies for mobile communications technology standards.

2. TECHNOLOGICAL BACKGROUND

The following description of basic mobile communications system architecture will provide a technological background with which to understand why and where technology standards are needed. The basic mobile communications system architecture consists of four main systems as shown in Figure 1 (Gallagher and Snyder, 1997).

1. **Radio Systems:** These systems include transceivers, antennas and controllers that are used to connect between a mobile device and a base station system through air-interface. Mobile device refers to cellular or mobile phones that end-users operate. The base station consists of the wireless transceiver and the transceiver controller equipment that serves one or more cells. It controls mobile communications functions on the network side when the mobile device communicates to the cellular networks (Gallagher and Snyder, 1997).

2. **Switching Systems:** These systems are interfaces within a cellular network and between the cellular network and other public switched telephone networks to coordinate the establishment of calls from and to a subscriber's mobile device. Public Switched Telephone Network (PSTN) refers to "the telecommunications network commonly accessed by ordinary telephones, key telephone systems, private branch exchange trunks, and data transmission equipment that provides service to the general public" (Gallagher and Snyder, 1997, pp. 412).

3. **Data-Based Systems:** Known as a location register, data-based systems provide information about subscribers along with service logic to control mobile services. These include a subscriber's phone number, the current location of the mobile device, subscribed features such as call forwarding and voice mail, and other options.

4. **Operations, Administration, and Maintenance (OA & M) Systems:** These systems are not shown in Figure 1, but they are integrated within systems to allow service providers to examine operations, to adapt the network equipment and functions, and to troubleshoot defects within the network.

Figure 1. Basic mobile communications system architecture (Source: Gallagher and Snyder, 1997, p. 11)

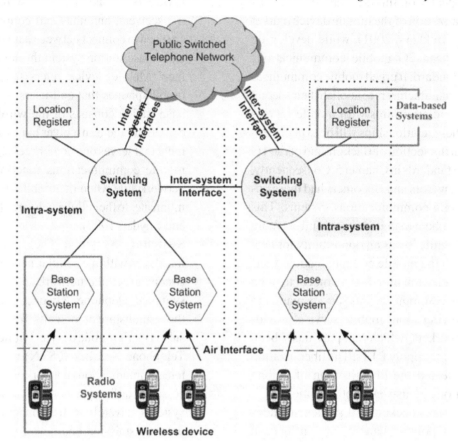

These systems need to be compatible and connected appropriately to function as a mobile communications network. A mobile communications network implies a set of systems built and owned by a mobile service provider to provide mobile communications service. There are three fundamental interfaces in the mobile communications system architecture (Bekkers, 2001; Mak, 2006).

1. **Air-Interface:** Refers to the interaction between mobile devices and base station systems. For example, a mobile device can't send or receive a signal if there is no base station that the device can reach or if the air-interface between the device and reachable base stations is incompatible due to different standards. For example, a mobile device embedded only with a CDMA chip

cannot be used in a GSM-based network due to the incompatible air-interface between the device and base stations.

2. **Intra-System Interface:** Refers to the interaction between various systems like base station systems and location registers within a switching system. For example, when two users subscribing to mobile communications service from the same service provider communicate with each other through their mobile devices, it requires only intra-system interface, because signals flow within the system.

3. **Inter-System Interface:** Refers to the interaction between two switching systems and between a switching system and other public switched networks. This type of interaction happens when there is communication

between a mobile device and a landline telephone, or between two mobile devices registered to two different mobile service providers. It is assumed that the two mobile service providers have a roaming agreement for the latter case. These cases require inter-system interface.

To provide service, a mobile communications network requires compatibility not only between components within one system, but also between different types of systems like base station systems and switching systems.

If there are industrially shared standards for these interfaces, a mobile service provider can purchase different types of systems from various manufacturers and integrate them horizontally to build a mobile communications network. If not, it needs to purchase all systems from one manu-facturer, which is a vertically integrated mobile communications network, because it is impossible for purchasers to figure out how the manufacturer integrate systems unless the manufacturer reveal its internal standards for systems integration. Vertical integration means that a single company (actor) controls the various steps in producing and/or distributing a product or service in order to increase its supply power in the market (Simchi-Levi, et al., 2003).

The existence of various mobile communications networks that are vertically integrated is not a problem if subscribers stay within the range of their network, because the vertically integrated networks tend to be incompatible with each other unless they share industrial standards. However, if subscribers travel beyond the range of their mobile communications network, then their mobile devices become useless in another mobile communications network.

The basic mobile communications system architecture (Figure 1) reveals where standards are needed for users to use their devices regardless of divided service territories by different mobile service providers. A subscriber can use her/his mobile device in another mobile communications network if the two networks have standards for air and inter-system interfaces and a roaming agreement. A roaming agreement is a consensus between mobile service providers to provide service that allows a subscriber to use her/his device in a mobile communications network other than the one (s)he subscribes to.

Standardization between incompatible mobile communications networks is a technologically complex matter because of the large numbers of different technologies used in the mobile communications network. In order to make two existing incompatible networks compatible, parts of a system or a whole system need to be changed, which can be extremely expensive or sometimes technically impossible. For these reasons, standards in the mobile communications industry are usually pre-defined by various actors before systems are developed or built. Since the creation of a mobile communications network requires a large number of technologies and great resources, usually many different actors are involved in this process – some that own necessary technologies and others that want to access the technologies with as low costs and barriers as possible. This social complexity among actors based on the technological complexity of mobile communications makes the standard-setting processes of mobile communications technologies very complicated.

Building on this background, the general categories of actors that can potentially be involved in standardizing mobile communications technologies is introduced in the next section.

3. GENERAL CATEGORIES OF ACTORS

To analyze organizational strategies for technology standards in the mobile communications industry, we need to identify the actors that play significant roles in this process. There are four different types of major actors.

1. **Mobile Service Providers:** Are the most significant players, because a technology will not survive without their commitment to adopt the technology. There are two kinds of mobile service providers. One is a government-owned operator, which was the most common kind of organization that provided mobile service in most regions except the U.S. before the liberalization of the communications markets. The other is a private operator, like companies based in the U.S. or other regions after the liberalization of the communications markets in the 1980s and 1990s. These two kinds of operators tend to behave differently. Government-owned companies, often run by the PTT (Post, Telegraphy, and Telephony) bureaus of governments, are mandated to act on behalf of their nation's interests rather than profitability, while private operators are similar to other corporations in that their main motives are profit. For example, France Telecom (former DGT), Deutsche Bundespost, and SIP used to be the French, West German, and Italian PTT, respectively, while Vodafone, Verizon, SK Telcom, and KDDI started as a private operators.

2. **Manufacturers:** Refer to companies that produce communications systems equipment used in networks or to maintain networks (e.g. switching systems, base station systems, location registers, mobile device, and other transmission equipment like cables and fibers). There are only a few large manufacturers that are able to manufacture all kinds of systems, because the cost and scale of the equipment needed to develop switching systems are high, so only a limited number of manufacturers have enough resources and capabilities to produce them. Thus, there are many more small companies that only participate in manufacturing mobile devices or other small components. For example, Ericsson, Alcatel and Siemens have focused

on networking equipment, while Motorola started as a car radio manufacturer and has since focused on mobile devices.

3. **Standard-Setting Bodies:** Are organizations where different actors come together to define standards for technologies. They can be formed by governments, research organizations or companies. Standards coming from these organizations are usually considered to be recommendations rather than regulation, so the successful implementation and diffusion of standards depend on the organizations' power to influence other potential adopters and the commitment of their members. However, it is also true that the recommendation can act more like an unspoken mandate in certain regions. Some examples of standard-setting bodies are: European Telecommunications Standardisation Institute (ETSI), founded in 1988; International Telecommunication Union (ITU), which is one of the specialized agencies of the United Nations; and 3rd Generation Partnership Project (3GPP), which is a collaboration agreement for the third generation mobile communications systems between ETSI, Association of Radio Industries and Businesses (ARIB)/ TTC (Japan), CCSA (China), ATIS (North America) and TTA (South Korea), founded in 1998.

4. **Governments:** Are the fourth major actor in standardization. They also involve themselves in the mobile communications industry by subsidizing research, allocating frequencies, issuing licenses, regulating companies, and sometimes, directly deciding on national standards for their regions. The power of their position can change over time as the industry evolves and can vary in different regions. For example, the Korean government played a significant role in implementing CDMA-based technology in Korea, and the Japanese government en-

couraged and coordinated the development of the Japanese standard for 2G mobile communications technology, PDC (Personal Digital Cellular). (Note: Korea in this book refers to South Korea.)

Unlike industries such as the computing machinery industry, end-users (consumers) do not play an important role in determining standards in the mobile communications industry, because end-users (consumers) simply use the mobile service that mobile service providers offer. A mobile service provider in the liberalized market is the main actor that decides which standard it will adopt. In many cases, end-users (consumers) do not even know what technological standard is embedded in their mobile devices.

These four categories of actors are inter-dependent on each other. Mobile service providers need a license from the government and mobile communications equipment, including handsets, from manufacturers. For manufacturers, mobile service providers are their primary customers. Although mobile device manufacturers can sell products directly to customers, they must cooperate with the operators to deliver feasible devices for the operators' systems. In many cases, it can be easier and faster for users to use the distribution channels of the operators. Standard-setting bodies are where different actor categories try to leverage their power to influence the standard-setting processes. These inter-twined relationships among actor categories have made the industry much more complicated, especially during the technology standardization process.

The roles and positions of these actors in the standardization of mobile communications technologies change with the evolution of technology standards. Actors compete, reconcile, or cooperate with each other depending on their interests. Even actors in the same category, for example, mobile service providers, do not always behave homogeneously because their interests can be conflicting.

However, these general actor categories can help to identify specific actors and their fundamental differences of interests. From this starting point, these actors' situations can be analyzed based on the existing self-organized configuration in the mobile communication industry and the networks that these actors try to form and shape.

With this general background of the mobile communications industry, the major actors and their strategies for the 1G mobile communications technology standards will be introduced. The global 1G mobile communications technology industry was too segmented by various standards for one standard to dominate. Among these standards, some standards became more important than others in influencing the industrial self-organized configuration during the 1G period and eventually influenced the industrial evolution of the 2G mobile communications industry. Thus, it is important to review the history of standard-setting processes of the 1G mobile communications technologies.

REFERENCES

Bekkers, R. (2001). *Mobile telecommunications standards: GSM, UMTS, TETRA, and ERMES.* Boston, MA: Artech House.

Gallagher, M. D., & Snyder, R. A. (1997). *Mobile telecommunications networking with IS-41.* New York: McGraw-Hill.

Mak, K. T. (2006). *Standard strategy for high-tech business.* Chicago, IL: University of Illinois at Chicago.

Noam, E. M. (1992). *Telecommunications in Europe.* New York: Oxford University Press.

Simchi-Levi, D., Kaminsky, P., & Simchi-Levi, E. (2003). *Designing and managing the supply chain: concepts, strategies, and case studies.* New York: McGraw-Hill.

Chapter 5
The 1G (First Generation) Mobile Communications Technology Standards

EXECUTIVE SUMMARY

During the development of the first generation (1G) mobile communications technologies, many organizations had not thought about standardizing a mobile communications technology. Due to the fact that a mobile communications market was regional and a network was run by a government agency or a monopoly organization, the standardization of a mobile communications technology was not occur to them. However, an interesting movement came from Nordic countries (Denmark, Finland, Iceland, Norway, and Sweden), which was standardizing a mobile communications technology. This chapter reveals the standardizations of the first generation (1G) mobile communications technologies including the Nordic Mobile Telephone (NMT) standard and the impact of the NMT standard on the history of the mobile communications technology standards.

1. BACKGROUND OF THE 1G TECHNOLOGY STANDARDS

The first generation (1G) mobile communications technologies had limited capacity, serving only niche markets for the military, certain government agencies and users in special industries (e.g. loggers, construction foremen, realtors and celebrities). In the 1960s and 1970s, this service was geographically limited and the mobile device

DOI: 10.4018/978-1-4666-4074-0.ch005

was too large, so it was usually mounted in cars or trucks; the smallest was a briefcase model. This form of mobile communications were not ready for mass development, because of (1) the limited capacity to service the general population, (2) the limited technology capability to cover large areas, (3) the large size of the mobile device, and (4) the high prices of mobile devices and tariffs.

In the 1970s, countries were still focused on building nation-wide landline communications network rather than mobile networks for a few customers. At least in developed countries, mobile service providers that were government-owned PTT (Post, Telegraphy, and Telephony) bureaus or monopoly companies like AT&T developed or adopted any available technologies to provide mobile services during 1960s and 1970s without considering technology standardization for potential future markets. Therefore, the existing self-organized configuration of the industry before the 1G was fragmented and dominated by monopoly PTTs or companies that had close relationships with governments (See Figure 1).

One important thing that differentiated the 1G mobile communications technologies from previous technologies was the cellular technology. The mobile communications technologies before the 1G era focused on developing a powerful base station system that could send signals as far as possible to cover the large area. The coverage of single base station was about 50 miles or more, which was enough to encompass most metropolitan regions at the time. Given frequency bands in a metropolitan area, the very limited number of subscribers had to use the mobile communications channels at the same time. For example, in the entire metropolitan area of New York City in

1976, only twelve channels served 543 subscribers. Thus, most users spent significant time waiting to get a channel (Garrard, 1998).

The cellular concept was a radical idea when it was formulated in 1947, because as mentioned, most research focused on transmitting a signal as far as possible, while the cellular concept proposed to deliberately limit the range of a signal transmission (Garrard, 1998). Each limited area would have one base station, which was called a "cell." In this way, the same frequencies used in one cell could be used in different cells, so the systems based on this concept would allow frequencies to be reused for more subscribers in a region (Garrard, 1998). This was how the name "cellular phone" came about.

It took many years to implement this idea, because the necessary technologies (e.g. electronic switches, integrated circuit for transistors, and handover technique when a user moves from one cell to another) were developed much later (Calhoun, 1988). The structure to support this concept became the basic mobile communications system architecture (Figure 1 in the Chapter 4) explained before. Thus, all systems mentioned in the architecture were needed.

Figure 1. The configuration before the 1G technologies standardizations

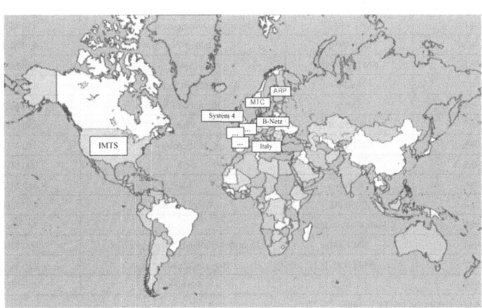

As emerging problems were resolved, the 1G mobile communications based on analog and cellular technologies became commercially viable during the 1980s in developed countries (e.g. Western European nations, Japan and U.S.). For example, a 1G mobile communications service was introduced in 1981 for the Swedish and Dutch markets, in 1982 for the Norwegian and Finnish markets, and in 1983 for the U.S. market. Table 1 shows the countries and their commercialized 1G standards with the year that service was first introduced.

In the beginning stages of commercial introduction of cellular technology in developed countries, the market remained very small, because of the still limited capacity of mobile communications systems and the high prices of devices and services.

Moreover, the 1G mobile communications networks were isolated because the mobile service providers (MSPs) and host countries developed their own standards that were incompatible with each other. For example, AT&T in the U.S. introduced AMPS (Advanced Mobile Phone System), which was developed by Bell Labs and operated in 800 MHz, while Scandinavian countries (Finland, Norway and Sweden) started to develop their own standard, NMT (Nordic Mobile Telephone) that was operated in 450 MHz. Germany, France, and Italy also developed their own standards, C-450 by Siemens, Radiocom-2000 by Matra, and RTMS by Italtel respectively (see Table 1). Italtel was also a government-owned firm. Consequently, by 1990 there were six different standards with eleven different variants used in Europe alone that were all incompatible with each other (Bekkers 2001).

The emerging configuration of the mobile communications industry during the 1G technology standardizations was not so different from the existing configuration before the 1G period. One reason for small changes in these configurations was that the previous industrial structure, rooted in the traditional relationships among the three major actors (government, mobile service provider and manufacturer), remained the same. This entrenched structure is described below.

Table 1. 1G technology standards adoption

Commercialized Year	Standard	Country	Frequency band
1979	NTT-AMPS	Japan	800 MHz
1981	NMT 450-25	Denmark, Saudi Arabia, Sweden	450 MHz
1982	NMT 450-25	Finland, Norway	450 MHz
1982	NMT 450-25S	Spain	450 MHz
1983	AMPS	AT&T Illinois Bell (U.S.A)	800 MHz
1984	AMPS	Republic of Korea	800 MHz
1985	NMT 450-20A	Austria	450 MHz
1985	NMT 450 B	Netherlands	450 MHz
1985	TACS	UK	900 MHz
1986	RTMS	Italy	450 MHz
1986	C450-20	West Germany	450 MHz
1986	RadioCom system	France	Multiple
1986	NMT900	Denmark, Finland, Norway, Sweden	900 MHz
1987	NMT 450 B	Belgium, Luxembourg	450 MHz
1988	C450-25	Portugal	450 MHz

(Source: Bekkers [2001] and UMTS Website)

During and after World War II, most countries nationalized their post, telegraphy, and telephony services, so that most mobile service providers were operated by each government's PTT (Post, Telegraphy, and Telephony) bureaus. This period has been called the "PTT era" (Bekkers, 2001). During this PTT era, governments and monopoly mobile service providers had a very close relationship and shared the same interests (Noam, 1992). As a major employer in a country, PTT ran its business reflecting national interests. For example, it purchased expensive communications equipment from its national manufacturers instead of companies headquartered in other nations, even if the latter offered cheaper prices (Noam, 1992). These national manufacturers such as Alcatel in France, Siemens in West German, and Italtel in Italy enjoyed constant procurement from the national PTT through a cartel and subsidy from governments to develop new innovative technologies for future mobile communications equipment. A cartel is an explicit agreement among sellers to share certain markets among them by fixing prices, allocating territories, etc. With the cartel, companies can maintain their market shares and manipulate the market. These traditional business practices among the three actors led many countries to develop their own 1G standard when they had the resources and capacity. As a result, the emerging configuration of the 1G mobile communications technology standards at the industry level was not drastically different from the existing configuration. Nevertheless, there were noticeable changes in terms of the emergence or demise of certain major actors. It is worthwhile to introduce the actors whose standards strategies influenced the evolution of the industrial configuration, and that would later play critical roles for setting the future 2G mobile communications technology standards.

In the arena of 1G mobile communications technology, different standards such as NMT, AMPS, C450, RadioCom and RTMS dominated their respective regions. It is necessary to sort out the actors that influenced the development and implementation of these different standards and figure out how these standards shaped the technology and market configurations.

The rest of this chapter is organized as (1) specific actors, that played a major role in standardizing a technology in their respective regions, (2) their situations, interpretations, and strategies for the 1G mobile communications technology standards, and (3) the impact of each standard will be will be identified, reviewed, and explained in the section of NMT (Nordic Mobile Telephone) Standard through the section of PTTS in France, West Germany, and Italy. Then, the actors' situations, interpretations, and strategies will be compared in the section of Comparison of Actors in the 1G Mobile Communications Technology Standards followed by assessing the emerging self-organized configuration during the 1G mobile communications technology standardizations in the section of Emerging Self-Organized Configuration: Market and Industrial Structure and the 1G Technology Standards.

2. NMT (NORDIC MOBILE TELEPHONE) STANDARD

2.1. Background of NMT Group

In the late 1960s and early 1970s, Nordic countries (Denmark, Finland, Iceland, Norway, and Sweden) planned and constructed national mobile telephone networks to serve commercial road traffic (Lehenkari and Miettinen, 2002). However, the increase of cross-traffic between these countries made it problematic for customers to use their devices in neighboring nations because their devices did not work in other regions. Mobile service providers from different nations not only did not have roaming agreement but also their systems were incompatible with each other at the time.

Acknowledging this problem, Swedish PTT suggested at the Nordic Teleconference in 1969 the possibility of building compatible mobile communications networks among the Nordic countries (Lehenkari and Miettinen, 2002), proposing both short-term and long-term plans. For the short-term plan, it asked that other Nordic countries adopt the manual mobile telephone, Mobile Telephone system D (MTD) that the Swedish PTT planned to introduce for its national network. For the long-term solution, it suggested cooperation among Nordic countries to develop an automatic mobile communications network together. It was not surprising that the proposal came from the Swedish PTT, because Sweden was historically the most influential country in the Nordic region and Swedish companies (e.g. Ericsson) had exported their products to the other Nordic nations.

Following the suggestion, the Nordic Teleconference formed a special working group called "Nordic Mobile Telephone Group (NMT group)" to research potential compatibility among existing mobile telephone networks and the possibilities of building future compatible automatic Nordic mobile communications networks (Lehenkari and Miettinen, 2002). The NMT group, except the representative of Iceland, convened regularly, produced reports for the teleconference, and defined and specified the requirements of systems beginning in 1970. Thus, the NMT group became the major actor in the regional standardization of NMT. Meanwhile, MTD was adopted in Denmark and Norway as the Swedish PTT had proposed. Finland, Denmark and Norway used a semi-duplex manual system, while Sweden had a full-duplex automatic system before the 1G technology standardization (Manninen, 2002).

The NMT group consisted of representatives from each Nordic PTT, which meant that the Nordic PTTs were not separate actors from the NMT group at least in the development phase of the NMT standard, because the Nordic PTTs shared the common goal of developing and building compatible networks. The reason they could fully cooperate with each other was the fact that they respected

each others' monopoly over their own markets and used their limited resources to collectively work out a way to standardize regional mobile communications technology. In the implementation phase, the role of the NMT group decreased as a recommender, while each PTT was in charge of implementing the developed standard in its nation. In this phase, each PTT had its autonomy to select and contract with its preferred manufacturers for the systems procurements.

The Nordic governments were not separate actors from the NMT group either, because the Nordic PTTs were a government department and run by government officials. Therefore the representatives in the NMT group represented each government's interests.

Another actor from the general actor categories that contributed the standardization of NMT was a manufacturer group. However, the role of manufacturers was rather passive. They were led by the NMT group in the development phase, and by each Nordic PTT in the implementation phase. In the development phase, the NMT group needed help from system manufacturers to solve technological problems, so it invited them in and outsourced to them to research technological problems. As a result, some Nordic manufacturing companies had the opportunity to be involved in the standardization of NMT at an early stage. For example, LM Ericsson, a Swedish company, was asked to introduce its mobile telephone system design to the NMT group as early as in 1970 (Lehenkari and Miettinen, 2002). However, the NMT group did not want to be locked into a few manufacturers, so it made the technology specifications publicly accessible to all manufacturers in the world. In the implementation phase, each PTT had the right to select manufacturers for systems procurements and installment of the standardized technology. Because of the common regional standard, the PTTs were not beholden to the manufacturers.

Thus, the major actors were the NMT group in the development phase and PTTs in the implementation phase. The manufacturers were minor actors to support the major actors. For the NMT standard, the development phase was critical, because the PTTs had already committed to implement the technology that the NMT group standardized. For this reason, the NMT group was the most important actor in standardizing the NMT. Thus, the NMT group's situation, interpretation, and strategies will be reviewed. Some of the important manufacturers that gained great capabilities and market positions during this process will be mentioned, because they became significant actors and took a leadership role in future 2G mobile communications technology standardization.

The situation and interpretation of the NMT group to be presented were about the time when the NMT group was formed in 1969.

2.2. The NMT Group's Situation

Organization's Capabilities to Meet Market Needs and Opportunities

The NMT group was formed to standardize a mobile communication technology among Nordic countries based on the practical needs of the market. Nordic countries had been cooperating with each other for economic development. Economic cooperation increased when Finland and Iceland joined the Scandinavian cooperation that had been established during the 19th century (Manninen, 2002). This cooperation increased trade among Nordic countries and resulted in having vehicles (including trucks carrying goods) crossing their borders freely. Incompatible mobile communications systems among Nordic countries were problematic when traffic became heavier across their borders, because mobile devices, usually attached to vehicles due to their size and technological limits, became useless in neighboring countries. Thus, there was a market need in the Nordic countries to have compatible mobile communications systems, leading to the creation of the NMT group. By nature, the NMT group was well aware of the

market needs and opportunities. Although it did not manufacture systems, it had the monopolistic power of the PTTs to leverage manufacturers in developing and standardizing a technology. As a representative group of the Nordic PTTs, it also had the capacity to invite other actors and form a value network. Therefore, it had capabilities to meet market needs and opportunities.

The Availability of Complementary Products or Compatibility of Products/Services

At that time, mobile communications technologies were not as sophisticated and complexly structured as technologies in the 2000s. Thus, this element, "the availability of complementary products" did not carry the same sense as it does for today's communications or computing systems. In the case of mobile communications, this element meant whether there were existing systems and technologies that could be compatible and supportable to the technology being developed.

From this perspective, the NMT group first reviewed existing communications systems whether they could be compatible. However, it realized that it was impossible to find compatibility among the existing systems. Therefore, the group decided to develop and implement a new common system that included radio, base stations, and switching systems to provide seamless voice delivery service over borders (Manninen, 2002). This meant that there were no existing complementary products and the group needed to develop all systems. This was why the Swedish PTT predicted ten years to develop and implement whole systems based on the NMT standard (Manninen, 2002).

The Type of Technology Innovation

Before and during the discussion of 1G technology standardization in the 1960s and early 1970s, all Nordic countries used manual mobile communications systems (Manninen, 2002). Denmark used System A and Norway used a similar kind of system (Manninen, 2002). They later adopted the Swedish manual system, MTD, in the early 1970s. The MTD system was first introduced in the Swedish market in 1971. Finland started to use ARP (Autoradiopuhelin) system in 1971 as well. The difference between manual and automatic systems was human interaction. The manual system required human operators to connect one user to another user, while the automatic system used mechanical switching that did not require human workers (Manninen, 2002).

The automatic system utilized advanced technologies to support duplex and handover techniques through new technologies like the microprocessor once it was available. This kind of the automatic system was in some ways the prototype of the proposed NMT. This implies that this part of the NMT technology was a continuous innovation in using integrated circuit technology and connecting to Public Switched Telephone Network (PSTN). However, the NMT technology was based on the cellular structure that had important features such as roaming. This meant that the NMT system required disruptive innovation for a new inter-system interface (for example, handover) when a mobile device moved one switching system to another. Thus, not only were the existing resources and capabilities used but also new resources and capabilities were required to develop the NMT standard (Manninen, 2002).

The Position of an Organization in the Market

The NMT group was formed by representatives of PTTs in all Nordic countries except Iceland at the beginning. Thus, the decisions of the NMT group reflected the shared regional aspirations of these PTTs that already had a great monopoly position in their respective markets (Manninen, 2002). Due to this fact, the position of the NMT group in the Nordic region was strong, because the participating PTTs were collectively commit-

ting to implement the standard the NMT group proposed and developed. Although the NMT group as a voluntary standard-setting organization did not have authority to force governments and other mobile service providers to adopt the technology, the agreement of the dominant PTTs effectively imposed an inescapable standard (Manninen, 2002).

The Availability of Alternative or Substitutable Technologies

When the NMT group began to discuss possible technologies for the regional mobile communications standard, cellular mobile telephony was not commercially available. When the NMT group researched existing alternative technologies in 1974, it found that the available options were all based on pre-1G technologies – for example, the automatic systems such as the Netz-B system launched in Germany and the Improved Mobile Telephone Service (IMTS) in the U.S. (Manninen, 2002). These technologies did not meet the needs of the NMT group in terms of calling channel usage, traffic capacity, and roaming capability (Manninen, 2002). The first cellular-based technology, AMPS, came later, but it did not meet the NMT group's needs as well in roaming. Thus, it was fair to say that there were no alternative or substitutable technologies when the NMT group reviewed the existing technologies.

Ownership and Deployment of IP (Intellectual Property)

The NMT group as a newly formed standard-setting organization did not have and did not intend to hold any IPR. Some of its members might have IPR, but the NMT group did not bother to distinguish who had essential or non-essential IPR ownerships. This was because the technological specifications had not been defined in 1969 and early 1970s, and perhaps because the NMT group did not foresee the relevance of IPR.

2.3. The NMT Group's Interpretation of its Situation

The Nordic PTTs perceived rising market needs and opportunities in mobile communications when traffic between Nordic nations grew rapidly along with the economic integration in the late 1960s and early 1970s. They interpreted that they could generate significant revenue by providing mobile communication services across borders. To provide these services, they needed to make their mobile communications systems compatible or require all to share a common system. In this situation, the NMT group was formed to represent these PTTs' collective and parallel interests and concerns. The group would administer necessary tasks for the Nordic PTTs to have a regionally standardized mobile communications system.

In interpreting the situation to develop a technology standard, the NMT group put high value on (1) opening up another source of revenue for the PTTs by providing the service far beyond their regions without jeopardizing their monopoly positions, (2) sharing the development costs for future cellular automatic mobile communications technology which would have been too great for one organization to afford, and (3) attracting more manufacturers to participate in developing and manufacturing necessary systems by providing to the whole Nordic market instead of a smaller national market (Manninen, 2002).

These three points illustrate the highly pragmatic and market-based value perspective of the Nordic PTTs. These Nordic countries shared similar needs and a common goal and had a history of interaction that surely helped the PTTs to cooperate rather than compete. Also, with this pragmatic perspective, the NMT group evaluated the compatibility of the existing systems and other existing technologies, found that it was impossible to make the existing systems compatible, and that the existing technologies at the time did not satisfy their needs, thus, decided to develop a new technology standard.

2.4. The NMT Group's Strategies

Configuration of Value Network

The Nordic PTTs (the members of the NMT group) were government-owned operators mandated to provide communications service to customers. Although some of them had a government-owned sister company manufacturing mobile communications systems (e.g. Telia in Sweden and Televa in Finland), the PTTs in the NMT group knew at the beginning that they had limited capability to develop all the necessary systems for the 1G mobile communications technology standard. For this reason, the NMT group needed help from manufacturers to configure a complete value network. Even at the early standardization stage, the NMT group was aware of this issue and invited various manufacturers to participate in the research. For example, the NMT group employed a Danish radio equipment manufacturer, Storno A/S, to work on signaling from 1971 to 1973 (Lehenkari and Miettinen, 2002; Manninen, 2002).

The NMT group also recognized the danger of the lock-in effect if it dealt with only a few manufacturers. To avoid being locked in, and to configure and secure a strong value network, the group strategically opened the technology specifications of the standard to all manufacturers around the world and encouraged many manufacturers to bid. Through this strategy, it could secure necessary systems and equipment for the 1G mobile communications service with as low prices as possible. Therefore, its strategy in configuring a value network was to select manufacturers through a competitive bidding process to form a horizontal value network with manufacturers.

Formation of Standard Setting

Although the Swedish suggestion of developing the Nordic mobile communications technology standard seemed abrupt in 1969, the Nordic PTTs quickly recognized the need and accepted the suggestion positively. Based on their interpretation of the situation, it would be beneficial for them to form a Nordic standard-setting organization to develop the technology standard together, so they could share the cost and attract more manufacturers. By the nature of the NMT group formed by Nordic PTTs (mobile service providers), the group focused on developing potential services rather than the technology itself. The group first wrote service-oriented specifications. For example, according to MT Report to 1973 Teleconference, it said, "A phone call should be possible from a mobile station to any subscriber connection or another mobile station in the same or any other country" (Manninen, 2002, pp. 66). After setting the service-oriented specifications, they investigated possible technologies that could satisfy all the desired requirements for services.

The capabilities of the NMT group and PTTs in technologies to develop systems for the specifications were limited. Thus, the NMT group invited manufacturers to join the standard-setting process and steered them in researching and developing technologies that could support the specifications.

Openness of IPR (Intellectual Property Rights)

The service-oriented NMT group did not consider owning IPR on the technologies it researched and developed, because it focused on securing more manufacturers to produce systems and equipment using the standard so that Nordic PTTs could purchase the required systems at low prices. However, the group was aware of the fact that a manufacturer could leverage its IPR to block other manufacturers from developing or manufacturing systems. Thus, when the NMT group outsourced to other manufacturers in researching or developing technologies, these manufacturers were not allowed to file patents on any work sponsored by the group (Manninen, 2002). NMT group's IPR strategy was to open relevant IPR to all manufac-

turers, in order to not worry about locking into any one or few manufacturers, and encouraged more manufacturers to produce necessary systems.

2.5. Impact of the NMT Standard

The NMT standard penetrated markets faster than anyone expected. The markets included not only the Nordic countries but other countries such as Saudi Arabia, Spain and Austria. The reasons for this fast penetration were (1) the NMT standard was the first international mobile communications technology standard; (2) it was commercialized earlier than other 1G standards except the Japanese NTT AMPS; (3) it was developed based on the pragmatic concerns of mobile service providers. These pragmatic concerns included market needs and opportunities and low implementation and maintenance costs; and (4) it did not depend on a single manufacturing vendor, so operators were not concerned about being locked into a few manufacturers (Bekkers, 2001; Manninen, 2002).

The NMT group's strategies were grounded in its situation and pragmatic interpretation, which valued the market needs and opportunities. The Nordic countries perceived the problem when trade and traffic crossing over their borders increased, so they worked together to solve this pragmatic problem by developing and standardizing the mobile communications technology together through the NMT group.

However, its success changed the aspirations of the NMT group over time from having a common standard regionally among the Nordic countries, to making it an international standard by advocating it to other countries. During the period of more than ten years in the standardization process, the NMT group, Nordic governments, and manufacturing companies found more opportunities to expand the market for the NMT standard. In particular, it was directly beneficial for national manufacturers in the NMT nations such as Ericsson and Nokia to be able to sell their products to other countries that would adopt the NMT standard.

Although the NMT group insisted in providing equal opportunity to all manufacturers around the world, considering where and how it negotiated with potential vendors, we can see that it favored Nordic companies. For example, only three companies, Ericsson, Motorola and NEC, submitted their bids for switching systems by the deadline, December 1, 1977 (Manninen, 2002; see Table 2). However, the group had already started to negotiate with Ericsson about various issues of its product, even before the assessment of the three companies' products came out (Manninen, 2002). In fact, the assessment revealed that NEC's switching system was most reliable (Manninen, 2002). Despite this conclusion, the group kept communicating with Ericsson to help Ericsson to improve its products. Ericsson's product was eventually chosen by all Nordic PTTs for switching systems.

During the standardization process, Finland, that had a historically different background from other Nordic counties including a peculiar relationship with the Soviet Union, perceived another opportunity to boost its electronics industry by establishing and positioning strong government-owned electronics companies such as Televa (Palmberg, 2002). The limited resources and capabilities of each company in Finland made it natural for them to cooperate with others in developing resource-consuming NMT systems (Palmberg, 2002; Steinbock, 2001). Along with the government policy, Nokia as a newcomer could position itself quickly as a systems manufacturer through forming a joint-venture with other firms like Televa and Salora. Nokia eventually absorbed Televa and Salora. Finland and Finnish manufacturers like Nokia proactively took advantage of the success of the NMT standard and its regional openness and emerged a significant actor in the future evolution to 2G and 3G mobile communications standards.

In the communications industry, Swedish companies have traditionally played significant roles. Meanwhile Finnish companies grew rapidly by taking advantage of the standardization sponsored by the NMT with the Finnish government's support. On the other hand, other Nordic governments and companies did not take advantage as much as Swedish and Finnish companies did.

3. THE AMPS STANDARD IN THE USA MARKET

3.1. Background of AT&T

The U.S. market has tended to be open and encourage free competition. However, AT&T had virtual monopoly power and played the role of a PTT in the U.S. market until AT&T's divestiture in 1984, even though there were a thousand small local service providers (King and West, 2002). AT&T was vertically integrated, from researching inno-

Table 2. Manufacturers participating in developing the NMT standard systems

System	Manufacturers
Switching System (MTX)	Invited companies: Ericsson, Motorola, NEC, C. Itoh, Telefenno (Televa + Nokia)
	Participating companies for bidding by the deadline December 1, 1978 – Ericsson, Motorola, NEC
Base Station System	Participating companies for bidding by the deadline June 30, 1978: Salora-Nokia (Finland), Magnetic (Sweden), Mitsubishi (Japan), Magnetic-Televa (Sweden-Finland), NEC (Japan), C. Itoh&Co Ltd (Japan), Tadiran (Israel), TeKaDe (Germany), Radiosystem (Sweden)
Mobile Device	Participating companies meeting for specifications in January 1977: Motorola, Sonab, AP Radiotelefon, SRA, Storno, Martin Marietta, Newcomers - Televa, Matsushita (Panasonic), Simonse Elektro A/S, Salora, Mitsubishi, and NEC

(Source: Manninen, 2002)

vative technologies to manufacturing equipment and providing local and long-distance telephone services. Its research center, Bell Laboratories (Bell Labs), was an indisputable world leader in developing communications technologies. Its manufacturing subsidiary, Western Electric, was the largest telephone equipment manufacturer on the globe. AT&T was an invincible company in the world communications industry, in terms of technological advancement and market size.

AT&T introduced commercial mobile telephone service in 1946 based upon its experiences as a radio provider for the U.S. Army during World War II. Soon after it launched the service, it realized the capacity limits of the technology due to sharing a single set of radio frequencies in a large area. To solve this problem and add capacity, a researcher in AT&T Bell Labs invented the "cell" concept that allowed multiple uses of the same frequency channel in different cells through dividing the large area into many cells. Pursuing this concept and leveraging its enormous resources and capabilities, AT&T became the first company that developed the technology of cellular telephony. AT&T developed and demonstrated a prototype using its cell telephony technology to FCC officials in 1962 (King and West, 2002).

AT&T also developed the first 1G mobile communications system, the AMPS (Advanced Mobile Phone System) in the early 1970s. However, the commercialization of the AMPS standard was delayed until 1983, because the FCC held up frequency allocation, there were problems in licensing the technology, and distraction was caused from the breakup of AT&T (Bekkers, 2001; West, 2000). Due to the fact that AMPS was developed so early, the specification of the AMPS implementation was rather obsolete by the time it was commercialized. For example, AT&T did not consider roaming services, thus the AMPS specification did not provide the definition of inter-system interface. This was because AT&T thought that AMPS would serve a single service provider (AT&T itself) for a specific region (e.g. New York City), rather than supporting

multiple networks run by various operators (Bekkers, 2001; West, 1999).

AT&T's dominant position and commitment to build universal landline networks made it perceive mobile communications technology merely as an enhanced but marginal service for high-end commercial niche users in urban areas. It did not recognize the mobile communications market as a new potential market that could be exploited. So AT&T kept focusing on building landline networks and neglected the mobile communications market. For example, it could have used the AMPS technology to connect customers in rural areas, or serve users traveling over larger areas, but instead it tried to reach customers in rural areas through landlines, and concentrated on serving mobile customers within a single urban region. One piece of evidence that AT&T did not pay enough attention to the mobile communications market was that it implemented the AMPS standard without considering the definition of inter-system interface, even though it had time to include this function before commercialization.

Nevertheless, AT&T was the actor that standardized AMPS in the U.S. market. Its situation and interpretation around the early 1980s when it commercialized 1G communications service based on AMPS technology will be reviewed. For the AMPS standard, this commercialization and diffusion period was more significant than the development period, because AMPS was born as a *de facto* standard. In the case of *de facto* standards, the commercialization and diffusion period is critical to standardize a technology in the sense of garnering market acceptance.

3.2. AT&T's Situation

Organization's Capabilities to Meet Market Needs and Opportunities

As the inventor of the cellular concept, AT&T was definitely capable of developing mobile communications systems based on AMPS technology or others to provide mobile communications service. As a monopoly service provider in the U.S., it was also able to standardize the technology it chose through its market power. As a vertically integrated company, it was capable of forming the necessary value network – with the exception of the mobile device manufacturing business, due to the Consent Decree signed in 1956 by AT&T and the Justice Department (Calhoun, 1988). This Consent Decree gave Motorola an opportunity to become a dominant mobile device supplier in the U.S. Thus, objectively, AT&T had the capabilities to meet market needs and exploit opportunities.

The Availability of Complementary Products or Compatibility of Products/Services

Although AT&T was able to reutilize the existing Improved Mobile Telephone Service (IMTS) systems for the AMPS system, it still needed to develop new systems based on new frequency bands and the radical cellular concept (Calhoun, 1988).

The Type of Technology Innovation

Although the cellular idea was a radical approach, the "cellular architecture was a system-level concept, essentially independent of radio technology" (Calhoun, 1988, pp. 39). So "it appealed to mobile system engineers, because it kept them on relatively familiar hardware ground" (Calhoun, 1988, pp. 39). AMPS systems themselves were not so different from the IMTS systems (Calhoun, 1988). Both technologies were based on Frequency Modulation (FM) transmission (Calhoun, 1988). Although AMPS systems required new functions such as handover, the AMPS technology was in part a continuous innovation based on existing capabilities and resources.

The Position of an Organization in the Market

AT&T had a unique position the U.S. Although the U.S. in general pursued open markets and encouraged free competition between companies, AT&T enjoyed an essentially monopolistic position, especially in domestic and international long-distance telephone service, due to the historical agreement between AT&T and the U.S. government. The agreement granted AT&T immunity against anti-trust prosecution in exchange for its commitment to establish universal telephone service in the U.S. (King and West, 2002). Under the auspices of this arrangement, AT&T grew fast and established great market position as a system manufacturer as well as service provider.

Although AT&T's monopoly was broken into seven independent Regional Bell Operating Companies after its settlement with the U.S. Department of Justice over the antitrust litigation in 1984, these regional companies inherited AT&T's technology standards.

The Availability of Alternative or Substitutable Technologies

When AT&T developed a prototype for the first generation cellular mobile communications technology in 1962, there were no alternative or substitutable technologies, because AT&T was technologically much more advanced than any other company in the world. However, during the very long interval from development to commercialization of AMPS (i.e. 1962–1983), NTT, the Japanese PTT, adapted the AMPS technology and released the NTT-AMPS standard, and the NMT group developed and commercialized its NMT standard. Thus, by the time AT&T launched its first commercial service in 1983, there were alternative technologies.

Ownership and Deployment of IP (Intellectual Property)

With AT&T founded on Alexander Graham Bell's patent, the company strongly acknowledged the importance of intellectual property rights early on its history and owned essential IPR on early telephony technologies. It invested in researching and developing innovative telephony technologies through its world famous laboratory and tried to gain IPR for these innovative technologies (Calhoun, 1988). While developing the cellular concept, Bell Labs filed a patent for the junction transistor that was needed to implement all the required features (Garrard, 1998). Moreover, AT&T owned various IPR essential to the AMPS technology through Bell Labs, for example, *U.S. Patent 3,663,762* (Cellular Mobile Communication System) issued May 16, 1972.

3.3. AT&T's Interpretation of its Situation

With its commitment to universal service – "making available, so far as possible, to all the people of the mandate for universal service..." (King and West, 2002, pp. 194) – which was a mandate based on the Communications Act of 1934s, AT&T succeeded greatly in establishing and dominating landline communications. For example, when the concept of cellular phone was invented in 1947, AT&T ruled 83% of U.S. telephones, 91% of U.S. telephone plants, and 98% of long-distance lines (Farley, 2008; King and West, 2002). Although AT&T could have used mobile communications technology to connect subscribers in rural areas to decrease costs when it was expanding its landline infrastructure to rural areas, it chose not to. It regarded mobile communications as a niche technology to provide special services for high-end commercial clients in urban areas. It regarded mobile communications as separate from its core mission, which was to provide universal landline telephony service.

Although AT&T objectively had the capabilities to meet emerging market needs and opportunities, it subjectively interpreted its situation and mission as unchanged – holding on to the monopoly and focusing on universal landline business (King and West, 2002). Thus, AT&T locked itself in a so-called competency trap. The competency trap phenomenon arises when an organization keeps concentrating on its existing services or products by applying even innovative technologies to maintain its existing systems or infrastructures rather than using innovative technologies for more creative services or products. This concept is investigated by Christensen in his book, "The Innovator's Dilemma" (1997). According to Christensen, incumbents tend to apply innovative technologies to improve their existing systems, structure, and market. In this sense, AT&T valued more the improvement of existing landlines and markets rather than potential market opportunities and needs.

3.4. AT&T's Standards Strategies

Configuration of Value Network

It seems that AT&T did not have specific strategies in standardizing the AMPS technology. Instead, it focused on its landline business. Thus, it just used its existing capabilities and vertical value network to develop and implement the AMPS technology. Not perceiving the mobile communications market as important, AT&T was not concerned at all in establishing and owning mobile communications technology standards in this emerging 1G market. In fact, it shrunk the business sector for mobile communications to one department.

Formation of Standard-Setting

With its monopoly position in the U.S. market and strong capabilities over the value network, AT&T did not see any need to form a standard-setting organization, so it developed and implemented the AMPS technology standard by itself.

Openness of IPR (Intellectual Property Rights)

AT&T was a technology innovator in the communications industry at the time. As mentioned, AT&T had been aware of the importance of IPR ever since it was founded. As many vertically integrated companies did, AT&T researched and developed technologies, closed its IPR, and manufactured systems using its proprietary technologies. However, it was forced to give up its mobile device manufacturing sector, so it had to rely on other mobile device manufacturers like Motorola. AT&T welcomed competition among mobile device manufacturers, because it did not want to depend on one or a few mobile device manufacturers (Calhoun, 1988). AT&T did not have to open its IPR for others to manufacture mobile devices based on AMPS, because FM (Frequency Modulation) was not a proprietary technology (Calhoun, 1988).

3.5. Impact of the AMPS Standard

First of all, AT&T did not pay enough attention to the upcoming mobile communications market. Thus, its strategies were driven by its existing systems, technologies and markets. In addition, the AMPS technology lacked some functions like roaming and security when it was commercially available. Nonetheless, the AMPS standard was successful in terms of diffusing over a large number of subscribers. One of reasons for its success was that the U.S. as a single country had the largest market for mobile communications service at the time and there was no competing technology in its territory. Owning the large market meant that AT&T could gain economies of scale faster and more easily than others, even though the growth of subscriber numbers was slower than that in smaller markets like Sweden and Finland. Economies of scale could decrease the prices of AMPS systems. The lower prices made AMPS

more competitive for countries that had planned to adopt the technology. In addition, AT&T, as a world leader, had been providing communications systems to other countries that had politically close ties with the U.S. government. For these reasons, AMPS was adopted in many countries in Asia and the Americas.

In fact, the first commercialized standard, NTT-AMPS in Japan, was an AMPS variant. Another 1G technology standard that had the largest number of subscribers at one point in Europe was Total Access Communication System (TACS), which was another variant of AMPS adapted in the U.K.

Many researchers believe that AT&T's breakup delayed the evolution of the U.S. mobile communications market. Although the breakup might have delayed market development, it did not affect the standardization of the AMPS technology as much as speculated. Indeed, it was AT&T's competency trap that shaped its standards strategy more than the external factors (e.g. settlement of the antitrust litigation). Although there was enough time between the development and the commercialization of AMPS, AT&T did not include newer and more useful functions like roaming.

In fact, in the long run, AT&T's breakup positively stimulated competition between service providers to move the industry forward. Some service providers split from the original AT&T and some newly emerged companies had the opportunity to choose a technology different from AT&T when they later moved on to the 2G era (choosing CDMA-based technology). Meanwhile, other companies depended on the path from existing AMPS systems, so these companies developed D-AMPS (Digital AMPS) for 2G mobile communications technology, which was continuous innovation. Without the AT&T breakup, this monopoly company would have simply adopted the continuous innovation of AMPS. In this perspective, the AT&T breakup expedited the industrial and technology evolutions through competition.

4. PTTs IN FRANCE, WEST GERMANY, AND ITALY

4.1. Background of PTTs in France, West Germany, and Italy

In Europe, France, West Germany and Italy were three countries that independently developed their own 1G mobile communications technology standards. They each had sizeable markets in terms of population. In addition to these similarities, their situations and interpretations for standards strategies were very similar, especially between France and West Germany. Thus, the three actors will be considered together in this section. (It was still the cold war period when developed European countries (mainly Western European nations) tried to develop 1G mobile communications technology standards. Therefore, "Europe" in this book refers to Western Europe.)

All three countries had large population (see Table 3) and their national manufacturers had capabilities in developing communications technologies, particularly in the case of France and West Germany. For example, the major French and West German manufacturers, Alcaltel/CGE (including former ITT) and Siemens respectively, ranked second and third in communications-related sales in the 1980s, while Italian manufacturer Italtel/STET ranked thirteenth or fourteenth (Bekkers, 2001, pp. 55). The larger population represented

a larger market size. Economically, a large market size favors the development and implementation of capital-intensive systems like communications. Companies tend to invest their resource only when they perceive that Return On Investment (ROI) will be positive, not negative. A larger market offers better prospects for positive and higher ROI than a smaller market.

The markets of these three countries were also controlled by their respective monopoly PTTs (Direction Générale des Télécommunications [DGT] in France, Deutsche Bundespost [DBP] in West Germany, and SIP in Italy). These PTTs had power over decisions about 1G mobile communications technology standard for their nations. The PTTs' situations – (1) monopoly positions in their large domestic markets and (2) strong ties with national manufacturers that had the capabilities to develop a technology – influenced PTTs to develop their own standards, even though there were alternative technologies. As a result, they adopted different standards – RadioCom 2000 system in France, C450 (also called Netz-C) in West Germany, and RTMS in Italy (see Table 4).

The PTTs outsourced the technology development to their national manufacturers. For example, DGT (now France Telecom) outsourced to Matra for the RadioCom system; DBP (German PTT) outsourced to Siemens for C-450; and SIP (Italian PTT) outsource to Italtel for its standard, RTMS (see Table 4).

Table 3. Population by country

Country	Population	Country	Population
Denmark	5,117,000	*Finland*	4,764,000
Norway	4,073,000	*Sweden*	8,294,000
Total of four Nordic countries		22,248,000	
France	53,478,000	*West Germany*	61,337,000
Italy	56,909,000	*U.K.*	55,883,000
United States	220,584,000		

(U.N. Statistics, 1981)

Table 4. Standards of France, West Germany, and Italy

Commercialized Year	Country	PTT	Developer	Standard
1981, 1982	Nordic countries	NMT group (coalition of Nordic PTTs)	More than 18 manu-facturers	NMT
1986	France	Direction Générale des Télécom-munications	Matra	RadioCom 2000
1986	West Germany	Deutsche Bundespost	Siemens	C450-20 (Netz-C)
1986	Italy	SIP	Italtel	RTMS

(Source: Bekkers, 2001; Manninen, 2002)

In contrast to the PTTs' exclusive outsourcing to a single manufacturer, more than eighteen manufacturers participated in the NMT standardization. The NMT standard was commercialized in 1981 and the AMPS standard was commercialized in 1983, while the Matra, RadioCom and C-450 were commercialized in 1986.

The following sections will show the analysis of the situations and the interpretations of the three European PTTs in the early 1980s when they began to pursue their standards strategies.

4.2. The PTTs' Situations

Organization's Capabilities to Meet Market Needs and Opportunities

The French, German, and Italian communications markets were monopolized by PTTs. These PTTs had the power to select both the technology as well as suppliers of systems. They preferred their national manufacturers when selecting a supplier, under special arrangements, even though the products that the national manufacturers provided were not necessarily competitive in price compared to the products of foreign companies in many cases. End-customers did not have a choice except to accept or reject whatever services the PTTs provided. In this traditional business practice, the PTTs were less aware of market needs and opportunities, and more interested in maximizing return on investment by postponing the development and implementation of new generation equipment,

while enjoying their monopolistic position by keeping the arranged relationships with their national manufacturers.

The Availability of Complementary Products or Compatibility of Products/Services

Mobile communications systems were vertically integrated at the time. Thus, a manufacturer produced all mobile communications systems and provided the systems as a whole to a mobile service provider. When the manufacturer developed new systems, they tended to develop all systems and integrate them as one without considering compatibilities with other systems. This was especially true in developing the 1G mobile communications technologies, since the manufacturers needed to develop every part of the system because of the new cellular concept.

The Type of Technology Innovation

As in the cases of NMT and AMPS, the German technology C450 and the Italian RTMS were based on the cellular concept. The French RadioCom 2000 was not purely based on the cellular systems even though it claimed to be so – it was a combination of some cellular ideas and traditional radio transmission techniques (Garrard, 1998). Thus, all of them were continuous technology innovations, and existing value networks were used to develop all necessary systems.

The Position of an Organization in the Market

The PTTs' positions were strong and dominating, being monopolies in their respective markets at the time.

The Availability of Alternative or Substitutable Technologies

There were alternative technologies such as the NMT and AMPS standards.

Ownership and Deployment of IP (Intellectual Property)

Because the PTTs outsourced the development of the 1G mobile communications technologies, the RadioCom system, C450-20, and RTMS were developed by individual manufacturer, Matra, Siemens, and Italtel, respectively. As a result, these manufacturers owned the IPR for the technologies.

4.3. PTTs' Interpretations of the Situations

Although PTTs were supposed to be independent from their respective governments, governments were still influential with PTTs as the largest shareholder. Because of this fact, the interests of PTTs had to align with those of their governments – providing national communications service to everyone with at a certain quality and price and promoting national manufacturers (Noam, 1992). Thus, PTTs favored their national manufacturers when they purchased systems, even though they often claimed that they fairly and impartially selected suppliers (Noam, 1992; Garrard, 1998).

With the large markets in these three nations and the capabilities that their national manufacturers had, it was preferable to develop technologies for their domestic markets instead of adopting a technology from other nations.

The French, German, and Italian PTTs interpreted their situation from a traditional perspective. Although there were alternative technologies (e.g. AMPS, NTT AMPS, NMT), they valued more developing their own and therefore national technologies. In developing and implementing the technologies, they also valued their national manufacturers more than other foreign manufacturers.

This interpretation could be explained with the competency trap of King and West (2002) as well. Different from the NMT group's market-driven value interpretation but similar to AT&T's, their interpretation was based on their existing markets and value networks, in particular, their relationships with their national manufacturers. This was the existing competency-driven interpretation.

4.4. PTTs' Strategies

Configuration of Value Network

Since the French, German, and Italian PTTs were used to requesting their national manufacturers to develop and supply the systems they needed, the PTTs exclusively outsourced the 1G technology development to one of their national manufacturers without considering adopting any of the existing technologies. Their strategy was to utilize the existing national value networks to develop the 1G technology standards.

Formation of Standard Setting

The three PTTs did not form or participate in any standard-setting organization, because they had the power to decide which technology should be developed or adapted as a standard for their respective markets.

Openness of IPR (Intellectual Property Rights)

The PTTs did not pay enough attention to claiming ownership of IPR when they outsourced the technology development to manufacturers. One reason was that the PTTs could indirectly force the manufacturers to license to other suppliers if necessary. Thus, the manufacturers that developed technologies claimed ownership of IPR. The manufacturers usually held their IPR internally without licensing to others unless there was a request from their PTT, so they could control the manufacturing industry.

4.5. Impact of PTTs' Technologies

As illustrated, the French, German, and Italian PTTs' strategies for the 1G mobile communications technologies were not different from their existing landline businesses, including value networks that had been implemented since they became powerful monopolies.

The technologies of these three PTTs were ill-suited to be adopted as international standards. First, there was the lack of functionalities (e.g. handover and roaming) in the technologies compared to NMT and AMPS. Second, the price of products and services was higher than those of the other technologies. Third, the three countries were large in Europe but not as large as the U.S., which meant smaller economy of scale than in the U.S. Fourth, there was a small number of manufacturers that produced systems, which implied the possibility of locking-in to a few manufacturers.

For these reasons, RadioCom 2000, C-450, and RTMS were not adopted beyond their borders (Bekkers, 2001). They became rather isolated standards within their domestic markets. It also implied that the manufacturers that developed RadioCom 2000, C-450, and RTMS did not have the opportunity to grow by exporting their systems to other regions as Ericsson and Nokia did.

5. COMPARISON OF ACTORS IN THE 1G MOBILE COMMUNICATIONS TECHNOLOGY STANDARDS

5.1. Comparison of Actors' Situations (1G)

The most significant difference in the situations between the NMT group and other actors was the NMT group's capabilities to meet market needs and opportunities. By its nature, the NMT group was formed to develop a standard that fulfilled emerging market needs and potential opportunities, while other actors, as a monopoly in their respective markets, concentrated on the development of technology standards as an extension of their existing resources, capabilities, markets and value networks. Another difference is that the NMT group promoted open IPR, while the PTTs left IPR strategy in the hands of the manufacturers (see Table 5).

The similarities between AT&T and the three PTTs were that they were monopolies in their national markets and had vertical value networks to develop and implement mobile communications technologies and systems. Although AT&T and three PTTs decided to develop their own standards, there were differences between them. AT&T developed its own AMPS, because there were no alternative or substitutable technologies when it developed the 1G technology; on the other hand, the three PTTs could have adopted one of the alternative technologies such as AMPS and NMT and not developed their own. Another difference between them was that AT&T as a manufacturer was keener to hold onto IPR for technologies than the three PTTs.

Table 5. Comparison of actors' situations (1G)

Element \ Actor	NMT group	AT&T	French, German, and Italian PTTs
Organization's capabilities to meet market needs and opportunities	(D) The NMT group developed the standard based on market needs and opportunities.	(D) Actors were in competency trap instead of trying to focus on market needs and potential opportunities	
The availability of complementary products or compatibility of products/services	(S) Actors needed to develop not only complementary but also core systems for their standards.		
The type of technology innovation	(S) All standards were a mixture of disruptive and continuous technology innovations, even though the degree between disruptive and continuous was different. Some functions of the standards were different.		
Market position	(S) They all had strong positions in their respective markets. The NMT group consisted of representatives of the Nordic PTTs.		
The availability of alternative or substitutable technologies	(D) There were not alternative or substitutable technologies at the time when they developed the technologies.		(D) There were alternative or substitutable technologies such as AMPS, NTT AMPS and NMT.
Ownership and deployment of IP (Intellectual Property)	(D) The NMT group did not own IPR.	(D) AT&T owned IPR on AMPS-related technologies	(D) The PTTs did not own IPR, but the national manufacturers to which they outsourced held IPR. However, the PTTs had power to force the manufacturers to license IPR.

[(S) Similarity, (D) Difference]

5.2. Comparison of Actors' Interpretations (1G)

The differences among these actors (the NMT group, AT&T, and the PTTs in France, Germany, and Italy) in interpreting their situations are noticeable. Considering that the total population of the Nordic countries was less than half of the population in France, it was a smart move for the Nordic PTTs to develop the 1G technology together. In this way, the group was able to share costs and attract more manufacturers to participate in the standard-setting process and to produce systems. The more manufacturers there were, the less expensive systems would be.

Recognizing the situation of the increasing trade and traffic among the Nordic nations, the NMT group interpreted the situation to develop the technology standard to meet this market need. Following market needs, the group defined the market requirements before specifying the technological requirements and researching possible technologies. Then, the group formed a value network to

deliver the technology standard it needed. The NMT group's interpretation was pragmatic, driven by market needs and opportunities.

AT&T's interpretation of the situation was driven by its existing resources, capabilities, market position and value networks. Although AT&T was the first to develop the cellular concept and a prototype based on a cellular system, it focused on the landline business heavily and underestimated the possibilities of mobile communications industry. It valued more expanding universal landline infrastructure and service. It believed that mobile communications service for selected customers in urban areas was simply a peripheral service that AT&T could provide.

The French, Germany, and Italian PTTs' interpretations of their situations were defensive. There were already alternative or substitutable technologies, but they valued the development of their own technologies using the capabilities of their national manufacturers. Developing the technologies also provided the protection for their domestic markets from other foreign manufacturers. Their perceptions were similar to AT&T's in terms of valuing their existing resources, capabilities, markets and value networks.

5.3. Comparison of Actors' Strategies (1G)

The NMT group's strategies were different from the strategies of AT&T and the three PTTs based on the differences in situations and interpretations. However, the strategies of AT&T and the PTTs were very similar as their situations and interpretations were similar (see Table 6).

All in all, these major actors' situations and interpretations were locally formed, ignoring international possibilities. However, the emerging configuration was self-organized internationally. The next section discusses this point.

6. EMERGING SELF-ORGANIZED CONFIGURATION: MARKET AND INDUSTRIAL STRUCTURE AND THE 1G TECHNOLOGY STANDARDS

The emerging self-organized configuration during the 1G mobile communications technologies standardizations was still fragmented by the various incompatible standards like the existing configuration before the 1G arena (see Figure 2). However, the NMT and AMPS standards were able to diffuse outside of their domains.

Considering the market size and power of Nordic countries at the time, the NMT standard was successful in terms of the growth of subscribers in the Nordic market and penetration outside of the Nordic market, even though the group did not consider expanding the market outside of its domain from the beginning. This was because of the NMT group's strategies that were driven by market needs and opportunities. In this sense, the NMT group formed a strong network with other actors such as the Nordic PTTs and manufacturers in the NMT standardization. Since the NMT was the first international standard in the global mobile communications industry, the group's strategies were (1) to form the value network with as many competitive actors as possible, (2) to use various standard-setting processes (e.g. bidding), and (3) to open IPR. On the other hand, AT&T and the three PTTs' strategies were to use their existing value networks and business practices including IPR policies to develop and implement the technology standards. The NMT network was more attractive and competitive than the networks formed by the French, German and Italian PTTs,

Table 6. Comparison of actors' strategies (1G)

Aspect \ Actor	NMT group	AT&T	French, German, and Italian PTTs
Configuration of the value network	(D) The NMT group formed new group-centered value network by encouraging more actors to participate.	(S) They utilized the existing value network to develop and implement the technologies.	
Formation of standard-setting	(D) NMT group was formed as a standard-setting organization to standardize the NMT.	(S) No standard-setting organization – de facto standard	
Openness of IPR	(D) The NMT group made sure that the manufacturers, who they outsourced to develop technologies, could not claim IPR on these technologies. It also opened the technology specifications to all manufacturers.	(D) Except mobile devices and related IPR, AT&T wanted to manufacture all systems by itself as much as possible without licensing their patents to others.	(D) The PTTs let the manufacturers, who they outsourced to, hold IPR. However, they were able to influence the manufacturers to license to others if necessary.

[(S) Similarity, (D) Difference]

Figure 2. The configuration of 1G standards in mid-1980s (Note: this picture is simplified and does not show the standards adopted by every country)

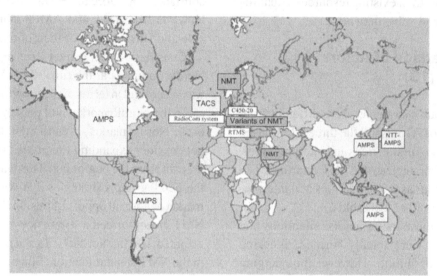

so the NMT network won over other networks in exporting a standard to other markets (e.g. Saudi Arabia, Belgium and Luxembourg).

Although AT&T's interpretation did not lead it to anticipate a bigger potential mobile communications market, its situation – as an inventor of AMPS, a leader in the communications industry, and holder of a virtual monopoly in the largest market in the world – made possible to diffuse its standard in many regions. Based on this analysis, AT&T had an advantageous situation compared to other actors, but its interpretation led it to not take full advantage of its situation. As a result, its strategies were just to utilize its existing assets and value networks. Nevertheless, the AMPS standard was successful, because AT&T's inherited assets (e.g. U.S. market size) and value networks (e.g. its world-leading manufacturing arm – Western Electric Company) were strong enough to diffuse its technology to other countries including Japan and the U.K. (e.g. King and West, 2002).

At the industry level, the configuration of the value networks that gradually emerged during the 1G mobile communications standardizations and development settled down and became the existing current configuration in the late 1980s when the 2G mobile communications technology standardizations began. The emerged configuration during the 1G mobile communications technology standardizations was still fragmented but there were differences from the one before the 1G period. The network formed by the NMT group played a significant role, while the networks formed by the French, German, and Italian PTTs became weaker and more isolated. AT&T maintained its network because of its inherited position that was embedded in its strategies for a 1G mobile communications technology.

REFERENCES

Bekkers, R. (2001). *Mobile telecommunications standards: GSM, UMTS, TETRA, and ERMES*. Boston, MA: Artech House.

Calhoun, G. (1988). *Digital cellular radio*. Norwood, MA: Artech House.

Farley, T. (2008). *Telephone history series*. Retrieved May 8, 2012, from http://www.privateline. com/TelephoneHistory/History1.htm

Garrard, G. A. (1998). *Cellular communications: Worldwide market development.* Boston, MA: Artech House.

King, J. L., & West, J. (2002). Ma Bell's orphan: US cellular telephony, 1947-1996. *Telecommunications Policy, 26*(3-4), 189–203. doi:10.1016/S0308-5961(02)00008-3.

Lehenkari, J., & Miettinen, R. (2002). Standardisation in the construction of a large technological system - The case of the Nordic mobile telephone system. *Telecommunications Policy, 26*(3-4), 109–127. doi:10.1016/S0308-5961(02)00004-6.

Manninen, A. T. (2002). *Elaboration of NMT and GSM standards: From idea to market.* (Academic Dissertation). University of Jyvaskyla, Jyvaskyla, Finland.

Noam, E. M. (1992). *Telecommunications in Europe.* New York: Oxford University Press.

Palmberg, C. (2002). Technological systems and competent procurers - The transformation of Nokia and the Finnish telecom industry revisited? *Telecommunications Policy, 26*(3-4), 129–148. doi:10.1016/S0308-5961(02)00005-8.

Steinbock, D. (2001). *The Nokia revolution: The story of an extraordinary company that transformed an industry.* New York: AMACOM.

West, J. (2000). Institutional constraints in the initial deployment of cellular telephone service on three continents. In Jakobs, K. (Ed.), *Information Technology Standards and Standardization: A Global Perspective* (pp. 198–221). Hershey, PA: Idea Group. doi:10.4018/978-1-878289-70-4.ch013.

Chapter 6
The 2G (Second Generation) Mobile Communications Technology Standards

EXECUTIVE SUMMARY

Based on the experience of the NMT standardization, some actors that had been involved in the NMT standardization (e.g., Ericsson) heavily participated in the standardization of a 2G mobile communications technology. They were successfully able to standardize the technology they pursued as the European mobile telecommunications technology standard. Considering all these European actors' early efforts, it was not a surprise that the GSM standard (the European 2G mobile communications technology standard) expanded its territory. The surprise came from the evolution of CDMA. How could the late entrant, Qualcomm, be able to create and expand the CDMA market? This chapter reveals not only fiery competitions among actors to standardize GSM and CDMA but also conflicts between GSM and CDMA camps as a group.

1. BACKGROUND OF THE 2G TECHNOLOGY STANDARDS

Mobile communications systems are much more complex than computing systems. The development and implementation of mobile communications systems require great resources. All systems need to be compatible and interoperable with each other to deliver signals seamlessly. For these reasons, the more the mobile communications market grows, the more technology standardization is necessary.

Many actors work towards potential market development. The nature of the projected market is influenced by currently available and future technologies. Thus, the actors also project specific usage and scope of a technology. Due to the co-dependent relationship between mobile communications technology and the market, they co-evolve. In this context, actors try to define the future market, choose a technology, and propose a competitive value network accordingly.

DOI: 10.4018/978-1-4666-4074-0.ch006

Moving to 2G from 1G mobile communications technology implied great changes in at least two ways. First, from the perspective of market evolution, the 1G mobile communications technologies were about creating a mobile communications market for niche users (e.g. high profile business people). Migration to the 2G market indicated that the mobile communications market was expanding for mass customers. Second, from the perspective of technology evolution, technological innovation was required in order to support the mass market. This chapter will discuss these two perspectives in depth, after which the general background of 2G technology standards will be presented.

1.1. Migration of Market from 1G to 2G

The demand for 1G mobile communications services exceeded the rate that anyone predicted. 1G technology was designed and developed for a small number of specific customers, for example, wealthy business people, celebrities, sailors, truck drivers, etc. Because of the large demand, this early market could have grown to include mass customers, but the problem was the technologies that the market was based on. The 1G technologies were limited in their capabilities and capacities to handle a large number of customers and therefore the limitations of the technology limited the growth of the market. Mobile service providers had to turn away potential new customers until there were other innovative technologies available to serve the mass market.

1.2. Migration of Technology from 1G to 2G

While the 1G market was becoming saturated, the digital revolution was happening and permeating into many industries. The rapid development in digital technology was astonishing. For example, the digital compact laser disk (the CD) was replacing tape cassettes and LP records in the entertainment industry. As digital technology evolved, prices of digital-related products (such as processors and memory devices) fell quickly. Although digital cellular and digital radio were not well understood in the mobile communications industry, digital technology was recognized as a key to solving the problems that the 1G market confronted (Calhoun, 1988).

The main technological problem was the limited capacity in 1G mobile communications systems. To solve this problem, actors needed to find a way to allow multiple users to share a frequency channel at the same time. There were various technologies such as Time Division Multiple Access (TDMA), Frequency Division Multiple Access (FDMA) and Code Division Multiple Access (CDMA) that multiplied access within a frequency band. These technologies required digital technology in order to be implemented and commercialized. For example, to burst signals into time, frequency, or code division channels, the systems had to be supported by digital technology.

One thing we should remember is that the timing between market and technology does not always match. Sometimes, as in this case, market development is delayed due to slow technology development or the unavailability of technology.

1.3. General Background of the 2G Technology Standards

On the top of these changes in technology and market development, there was another change in the socio-economic and political environments affecting mobile communications technology standardizations. Many countries started to liberalize their communications industries right before or during the period of 2G mobile communications technology standardization, which ruptured traditional business relationships and practices in the industry (Garrard, 1998).

As the European Union (EU) took shape, there was a general aspiration to strengthen EU as a whole. EU politicians recognized that it would be advantageous to have a single mobile communications technology standard in Europe, so they encouraged the industry to standardize a technology. In addition, the World Trade Organization (WTO) pressured nations to liberalize and open their mobile communications industries.

This was a challenging new situation for market-dominant organizations such as the PTT monopolies. These actors needed to interpret their new situations – that were also affected by the existing configuration – to develop their strategies for 2G mobile communications technology standards. These strategies interacted to form and shape reconfigured networks. The interactions of these networks eventually affected the evolution of the self-organized configuration.

In this political environment, the GSM group was founded for the purpose of setting a pan-European 2G mobile communications standard. The manufacturers, that had formed strong networks during the 1G mobile communications technology standardizations (e.g. Ericsson), were able to leverage their 1G success to promote their visions of 2G technology. Meanwhile, other actors, that had been unsuccessful in 1G such as the French, German and Italian PTTs, tried to regain their influence through the 2G standardization. We shall analyze how the standards strategies of these actors interacted and ultimately established the pan-European standard now known as GSM (Global System for Mobile communications, originally from *Groupe Spécial Mobile*).

The U.S. mobile communications industry at the time was going through great changes, including de-monopolizing AT&T and liberalizing the mobile communications industry. At the inception of AT&T's break-up in 1984, the Federal Communications Commission (FCC) granted two licenses in each market area to stimulate competition between operators. Operators, including the Baby Bells that came out of AT&T's breakup, adopted or inherited the AMPS technology for the 1G standard.

AT&T's breakup and the U.S. government's policies opened the U.S. mobile communications market to competition for 2G technology standards. Many mobile service providers (e.g. BellSouth Mobility, LLC, and U.S. Cellular) followed the natural migration path from 1G AMPS to D-AMPS (Digital AMPS) or TDMA (Time Division Multiple Access) for 2G mobile communications technology without exploring other possible technologies. However, a few others such as PacTel (now a part of Verizon Wireless) searched for new potential technology. This situation gave Qualcomm the opportunity to maneuver its strategies to persuade some operators to adopt its CDMA-based technology, which later impacted 3G technology standardization.

To understand how 2G standards have diffused over the world, we need to understand the role of actors in the rest of world other than Europe and the U.S.

The Korean government and Korean companies such as Samsung and LG that grew rapidly in the global electronics industry under OEM (Original Equipment Manufacturer) and ODM (Original Design Manufacturer) arrangements, were bystanders when foreign companies like Motorola and Ericsson came in and dominated the Korean mobile communications market during the 1G period. When it came time for 2G standardization, the Korean government perceived a chance to boost its electronics companies by adapting and commercializing the risky and only theoretically proven CDMA technology for the 2G mobile communications standard. Similar to how Nokia took advantage of the opportunity to build its capabilities and resources during NMT standardization, the Korean actors seized this opportunity to become major players in the global mobile communications industry through adopting the CDMA technology.

Japanese actors such as the Japanese government, operators and electronics companies, which had been the first to commercialize 1G technology, interpreted their situations to develop their own 2G standard, PDC (Personal Digital Cellular) with the ambition to standardize it internationally. However, they were not able to diffuse their standard beyond their nation. Therefore, they were isolated with their PDC standard and the Japanese manufacturers could not export their products during the 2G arena. Although the Japanese electronics companies achieved great success in the global market during the 1980s and 1990s, they were confined to their domestic market for mobile communications systems because of the technologically isolated standard.

In the rest of the world, countries adopted either GSM or D-AMPS in the mid-1990s since there was not enough commercial data to evaluate the potential of CDMA. Between GSM and D-AMPS, GSM was preferred due to its better functions (e.g. international roaming) and popularity in Europe. Once information about CDMA's viability and commercialization was available, some countries began to consider CDMA as a standard. Many countries already had the infrastructure for GSM or D-AMPS by then and it was too expensive to switch entirely to CDMA systems, but some countries such as China and India introduced CDMA as the second standard in their nations.

The emerging configuration during the 2G technology standardizations was greatly different from the existing configuration in the 1G arena. First, the fragmented mobile communications market with multiple standards in 1G became more unified under three standards globally (GSM, D-AMPS, and IS-95 [cdmaOne]) at the beginning and ultimately, under two standards (GSM and IS-95). Second, the mobile communications technologies became global, so actors had to choose between two standards (GSM and IS-95). The GSM technology as the first 2G standard had more time to spread than the IS-95 standard. However, the IS-95 standard was still able to penetrate some parts of the global market because of the smartly formed network by Qualcomm and the Korean government. Later, as 3G migration began to oc-

cur, CDMA technology was also chosen by some because of its superior technological capabilities for 3G migration.

Building on this brief background of the 2G technology standards, each major actor's standards strategies and the interactions of these strategies to form and shape networks will be reviewed. Then, the emergence of the industrial configuration through the interactions of the networks during the 2G mobile communications technology standardizations will be analyzed.

2. GSM STANDARD

In this section, first, a short background of the emergence of the GSM standard will be presented. Second, the situations, interpretations and strategies of three major actors (GSM group, French and West German governments, and Ericsson) will be presented separately. Third, these three actors' situations, interpretations, and strategies will be compared, then how they tried to form their value networks will be analyzed. Lastly, how the market and technology configuration of GSM emerged (or self-organized) out of the interactions of the value networks built by the strategies of the actors will be examined.

2.1. Overview of the Emergence of the GSM Standard

It was predictable that the 1G mobile communications market would be quickly saturated due to its limited capacity and coverage. Mobile service providers needed a next generation mobile communications technology that provided much more capacity with at an affordable price for the mass market. However, the development cost of a 2G mobile communications technology was higher than any one company could afford in Europe.

The PTTs in the Nordic countries that had benefited from developing the NMT standard saw an opportunity to develop a pan-European mobile communications technology standard. The European mobile communications market was at the time fragmented by the various 1G standards, so developing a pan-European standard would provide an opportunity for mobile service providers to share development costs and gain extra revenue by offering international calls through roaming, without competing against each other, while continuing to monopolize their domestic markets. In 1982, the Nordice PTTs suggested the idea of developing a pan-European 2G standard to European Conference of Postal and Communications Administrations (CEPT). CEPT, whose acronym came from its French name (Conférence Européenne des administrations des Postes et des Communications) was established in 1959 and consisted of PTTs from the European countries. CEPT accepted the suggestion of the Nordic PTTs and formed a working group called "Groupe Special Mobile," whose acronym, GSM, was applied to the technology, later re-named "Global System for Mobile Communications." This GSM group later became part of a standard-setting organization, the European Communications Standards Institute (ETSI), that was established in 1988. All through the 1980s, the GSM working group was an important actor as a leader in developing the pan-European 2G technology.

The development phase was significant for the GSM standard, because it was the time when all the technological specifications were decided. The actors involved in the GSM standardization focused their strategies on the development phase to gain competitive positions for the future. However, they understood that once they reconciled their conflicting interests and developed a common GSM technology, they would cooperate with each other to diffuse the technology broadly as a global standard.

Outside of the GSM working group, to be sketched below, three other actors (the West German and French governments and Ericsson) played important roles in shaping the GSM technology specifications.

The West German government considered the 2G mobile communications technology as an opportunity to boost its electronics industry. It witnessed how Nordic manufacturers like Ericsson penetrated and gained major share in the global market based on the resources and capabilities it acquired through NMT standardization, while German manufacturers were confined to the domestic market with its isolated 1G technology standard. So the German government decided to subsidize research conducted by its national companies like Standard Elektrik Lorenz (SEL) and Deutsche Bundespost for the 2G mobile communications technology from 1981 to 1983 in order to regain German influence (Bekkers, 2001).

The West German government also agreed with the French and Italian governments in 1985 about adopting the same technology that the GSM working group would support. A year later, the UK joined this agreement as well. This agreement from the four most populated countries in Europe guaranteed economies of scale for the mobile standard that the GSM group would select.

The German and French governments apparently believed that the GSM group would adopt the technology they supported. They therefore cooperated with each other in researching and advocating for technologies, and formed German/French consortia to subsidize them. An interesting point was that the largest national manufacturer in West Germany, Siemens, did not join any of these consortia because it was busy developing its 1G technology, C-450, at the time. This fact points to a weak link in the value network that the German and French government proposed. How the other actors exploited this weakness will be explained in the subsequent detailed analysis.

The Nordic actors had grown strong from successfully implementing the NMT standard. Therefore, it was natural for the Nordic actors to propose the idea of standardizing the pan-European mobile communications technology and try to influence and shape the GSM standardization.

The detailed analysis that follows shall make clear how the actors' standards strategies were formed and interactively resolved the competition between the German-French alliance and the rest of the GSM group in standardizing the pan-European mobile communications technology.

2.2. GSM Group

2.2.1. Background of GSM Group

Groupe Special Mobile within the CEPT organization was formed to research and standardize 2G mobile communications technology in Europe. In the 1950s, CEPT was established by nineteen European Post and Communication Administrations to promote cooperation among PTTs for postal and telecommunications services, in tune with the political movement to form the European Union among Western European countries (Bekkers, 2001). The activities of CEPT included administration, operation, commercialization, and technology standardization to provide homogeneous, coherent and efficient communications services by harmonizing and improving postal and telecommunications relations among the European countries (Bekkers, 2001; Toutan, 1985). Thus, the members of CEPT often conducted research for future technologies together and shared the costs.

The original members of CEPT were all monopoly PTTs that regulated as well as operated postal and communications services in their respective nations, so CEPT represented the interests of PTT members. Therefore, as one of the working groups within CEPT, the GSM group was under the influence of these members. Although CEPT contributed to harmonizing postal and telecommunications services in Europe, which was one of the EC's goals, its limitation in GSM membership (open only to PTTs) did not meet the requirements as an official standard-setting body (i.e. transparency and well-balanced representations from various types of actor groups like

manufacturers and users) (Bekkers, 2001). Under pressure from the EC and other actor groups like manufacturers, the GSM group was separated from CEPT and became part of the European Telecommunications Standards Institute (ETSI). It also opened its membership to other types of actor groups.

However, in 1987, a year before independence from CEPT, the GSM group made a significant decision about technology specifications that set the future technology direction for the GSM standard (see Figure 1 and detail in the section of Emergence of the GSM Standard and the European 2G Configuration through Actors' Interactions). This decision affected the future of major actors (the French and German governments as well as Ericsson), as well as the smaller actors that were later allowed to join GSM. Thus, the GSM collective as a group should be considered as an actor in the GSM standardization process. Indeed, GSM's goal and strategy set the context and constraints on the strategies of its members. Treating the GSM group as an actor, its situation, interpretation and strategy prior to the selection of the GSM standard in 1987 are presented below.

2.2.2. GSM Group's Situation

Organization's Capabilities to Meet Market Needs and Opportunities

Due to the limited capacity and high prices for services and products in the 1G mobile communications technology period, there was a need for a next generation technology that could provide larger capacity with lower prices than the 1G mobile communications technology standards.

Following the political unification movement in Europe, the GSM group was formed to create a pan-European mobile communications technology standard, not only to take advantage of the group's larger awareness of market needs, but more importantly to gain economy of scale and promote European technology. The GSM group itself did

Figure 1. The timeline of development of GSM standard

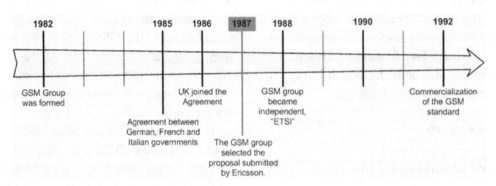

not have the capabilities to produce goods but it could leverage its implied power of the PTTs to orchestrate the technology development of the PTTs and manufacturers.

The Availability of Complementary Products or Compatibility of Products/Services

The TDMA, FDMA, or CDMA technologies that the GSM proposals were based on were different from the 1G mobile communications technologies such as AMPS and NMT, in that the new technologies were developed with the support of digital technology. Digital-based systems were incompatible with the existing analog-based systems. Thus, all systems needed to be developed.

The Type of Technology Innovation

The two distinguishable characteristics of the GSM specification were: (1) the switch from analog to digital technology; and (2) the interlining of different national networks. Chopping all continuous signals into a logical 1 (pulse present or high) and a logical 0 (pulse absent or low), which is called "digitizing," was disruptive innovation. The interlining of different national networks meant creating a seamless interface between switching systems (inter-system interface) requiring coherent communications and synchronization between systems. Thus, the technology for GSM was a disruptive innovation.

The Position of an Organization in the Market

As a volunteer organization that consisted of PTTs in Europe, it did not have the statutory power to interfere with each nation's sovereignty in developing or adopting a standard. However, its members, the PTTs, had monopoly positions in their respective nations. Thus, if its members agreed to adopt the technology that the GSM group supported, then the technology would be standardized because the support of the PTTs would guarantee the acceptance and diffusion of the technology. This position power of the GSM group, however, depended on the willingness of the PTTs to harmonize their technology and thereby abided by the political will of their governments to strengthen the EU.

The Availability of Alternative or Substitutable Technologies

When the suggestion for the pan-European standard came out in 1982, some nations had just commercialized and others were still developing or implementing the 1G mobile communications technology standards.

During the early development of the GSM technology from 1982 to 1985, the 1G market grew rapidly, especially in the nations that had adopted the NMT standard early on. The demand for mobile communications services began to exceed the capacity of the NMT systems in these nations. The countries that needed more capacity had two options: they could either wait until the GSM standard launched, or upgrade their existing systems for more capacity.

The mobile service providers in the Nordic nations and the Netherlands started to introduce another variant of NMT (called NMT-900) in 1986. The NMT-900 substantially improved capacity by utilizing higher frequency bands (900 MHz) over the existing NMT (450 MHz) standard systems (Bekkers, 2001). However, the problem was the frequency bands of 900 MHz, because these bands were reserved for the new GSM standard. Although NMT-900 was based on analog technology, some people considered it as an alternative technology for the GSM standard at the time, because it used the same frequency bands.

Outside the Nordic region, the growth of 1G was not as rapid. Therefore, Europe as a whole could wait for the pan-European standardization of 2G technology.

Ownership and Deployment of IP (Intellectual Property)

All components and systems related to the GSM technology had to be newly developed, whether they were essential or non-essential. Although the GSM group researched and tried to develop certain technologies, it needed cooperation from manufacturers that had the necessary technological capabilities and resources. Although the GSM group did not hold IPR, the configuration of IPR ownerships of essential and non-essential technologies would depend on its selection of proposals. This was one aspect where the GSM group had power over the manufacturers.

2.2.3. GSM Group's Interpretation of its Situation

Since the GSM group prior to 1987 was a collective of PTTs, its interpretation of the situation was driven by their perception of the market

needs and the potential profitability. The GSM group welcomed the idea of a pan-European mobile communications technology standard because the GSM group valued the possibilities 1) to share costs for the technology development, 2) to remove lock-in effects from manufacturers by having open interfaces between systems, 3) to reduce prices for infrastructure and mobile devices through economies of scale, and 4) to open up an extra source of revenue from international calls for PTTs by interconnecting national networks without competing against each other (Bekkers, 2001).

Indeed, the GSM group preferred any technology that could deliver these advantages rather than evaluating technology for technology's sake. One piece of evidence for the GSM group's market- and cost-driven set of values was that it delayed a technological decision between analog and digital systems until 1987, when most actors involved in technology proposals had already committed and begun to develop digital systems (Bekkers, 2001).

2.2.4. GSM Group's Strategies

According to its interpretation, the GSM group as a collective of PTTs pursued the following standards strategies.

Configuration of Value Network

The GSM group preferred open interfaces between systems, so it could decrease entry barriers for manufacturers that could not afford to develop all systems vertically to enter the industry. In the 1G arena, most standards had closed interfaces except the NMT standard, so a few large manufacturers that had enough resources and capabilities dominated the mobile communications system market. Having open interfaces would lower the entry barriers for new manufacturers, so there would be more competition among manufacturers. Moreover, if open interfaces were adopted as a standard, operators would not be locked into a few manufacturers and would have more options in purchasing systems at lower prices, which would happen through price competition among manufacturers. Thus, the GSM group chose to promote the configuration of a horizontal value network with open interfaces.

To promote such a value network composed of competitive actors, the GSM group encouraged various actors to propose prototype systems for the 2G mobile communications technology standard. The GSM group promised that it would compare and evaluate different prototype systems based on fair and objective criteria such as cost-effectiveness and openness of IPR, and would strive to harmonize the various proposals.

Formation of Standard-Setting

The GSM group set itself up as the venue for the pursuit of the pan-European mobile communications technology standard. Even after the GSM group became an entity independent of CEPT and a part of ETSI, it steadfastly suppressed the development of other technologies and venues that could possibly compete against the GSM standard in Europe (Bekkers, 2001).

Openness of IPR (Intellectual Property Rights)

The members of the GSM group, national PTTs in Europe, were the adopters of mobile communications technologies rather than developers. Collectively, they preferred to have open IPR as much as possible, so they had more choices for systems implementation without being locked into any manufacturer's products. The GSM group's strategy to promote open IPR was to encourage the actors who held IPR on technologies to license their IPR non-exclusively to all with fair, reasonable and non-discriminatory terms and conditions (so-called RAND licensing).

The GSM group discussed the possibility of patent pooling among patent holders, so the manufacturers that wanted to manufacture the GSM systems could quickly and easily license

necessary patents. However, the patent pool did not happen because of conflicting interests among the patent holders.

2.3. The West German and French Governments as Actors in 2G Standardization

2.3.1. Backgrounds of the West German and French Governments and their PTTs

The German communications industry had been based a close alliance between government and systems manufacturers (Noam, 1992). Although Bell's invention was introduced in Germany in 1877, it was not protected by patent in the German territory, so Siemens freely started to manufacture and introduce a German telephone system (Noam, 1992; Siemens, 1957).

Telephony systems in Germany were initially implemented in rural post offices, because the German Post perceived that telephony as a complement to state telegraphy could be used in rural areas where it was economically not feasible to have a trained telegraph operator (Noam, 1992). Later, telephony development was more focused in cities. For instance, although the difference of overall telephone density between the U.S. and Germany was quite large in 1902 (one per thirty-four persons in the U.S. and one per one-hundred-twenty-eight persons in Germany), the difference was much smaller just comparing cities. The numbers of people per telephone were thirty-nine in New York City and thirty in Chicago, while the numbers were forty-three in Berlin and thirty in Stuttgart. However, for larger regions that included rural areas, there were thirty-one people per telephone in New York State and twenty-four in Illinois, compared to six-hundred-thirty-four people per telephone in one eastern Prussian district and a little less than five-hundred in many regions in Germany (Holcombe, 1911; Noam, 1992).

Another characteristic of the German communications industry was that it was more export-oriented in the early 1980s than other nations in Europe except the Netherlands and Sweden, with exports reaching almost 30% percent of sales. In 1983, Germany was one of the largest exporting nations in communications products in the global market according to the electronics industry's trade association (Noam, 1992). However, the German government believed that overall the German information technology industry was falling behind in the world market, and needed to be redirected to develop and export next generation high-technology related goods (Noam, 1992).

This brief historical background of the German communications industry largely shaped the German government's perspective in standardizing the 2G mobile communications technology. Next, a brief sketch of the French communications industry will be presented.

The French government had a long history of intervention in the communications industry. From the early twentieth century to the mid-1970s, the French landline communications system had greatly deteriorated because of inadequate investment. This sorry state of affairs became a major campaign issue in the French presidential election in 1974 (Noam, 1992). After Valéry Giscard d'Estaing won the presidency, the French government established a five-year plan to modernize communications infrastructure, utilizing this modernization of communications infrastructure to boost the French high-technology industry to become competitive in the global market (Noam, 1992).

The government-owned PTT, Direction Generale de Telecommunications (DGT) – renamed France Telecom in 1988 – as a monopoly operator had exclusive relationships with its system suppliers including foreign firms such as the American firm ITT (International Telephone and Telegraph) and the Swedish firm Ericsson. However, having a relationship with foreign firms was against the French government goal of minimizing foreign competition in its domestic electronics market (Noam, 1992). Thus, the government encouraged French companies such as Thomson-CSF to acquire these foreign companies' shares in their French subsidiaries, while pressuring the foreign companies to sell their shares.

After a socialist government acquired power in 1981, the French government furthered this French-centric policy in communications industry through nationalization of the electronics sector (Noam, 1992). In the 1980s, the French government nationalized three of the five largest manufacturers that shared about 90% of the French communications market: CGE (CIT-Alcatel), Thomson (Thomson-CSF and LMT), and CGCT (Noam, 1992).

These background facts showed how comprehensively the French government was inclined to control its communications industry. This tendency largely shaped the French government strategies when it involved itself in the 2G mobile communications technology standardization process in the 1980s.

We now consider the German and French governments as actors in setting the GSM standard. Their situation, interpretation of the situation, and strategy prior to the critical 1987 GSM standard adoption will be presented below.

2.3.2. West German and French Governments' Situation

Organization's Capabilities to Meet Market Needs and Opportunities

As the 1G mobile communications systems were running out of capacity in many regions, the German and French governments as actors that virtually owned their national PTTs understood the market needs – namely, a new technology to generate more capacity.

Although they did not directly have the capabilities to set up networks and provide services, they had the power to align their national manufacturers and operators to develop and standardize a technology to meet their national market needs. Moreover, as the two largest nations in Europe, they had the capability and opportunity to shape the pan-European 2G standard.

The Position of an Organization in the Market

Both governments had great power to pressure their PTTs and national manufacturers in their jurisdiction. Their power was also related to their market size. West Germany and France were two of the most populated countries in West Europe. Thus, their support of the GSM group's choice of technology was necessary for the realization of a pan-European 2G mobile communications standard.

For the remaining four aspects of the situation they faced, namely,

- The availability of complementary products or compatibility of products/services.
- The type of technology innovation.
- The availability of alternative or substitutable technologies.
- Ownership and deployment of IP (Intellectual Property).

The West German and French governments essentially faced the same facts as the GSM group.

2.3.3. West German and French Governments' Interpretation of the Situation

The German and French governments' interpretations of the situations were driven by their political aspiration to boost their electronics industries (Bekkers, 2001). They saw how Ericsson from Sweden (a much less populated and politically less significant country), based on its successful experiences with the NMT standard, expanded its market not only in Europe but also in the U.S. and other regions and became a major mobile commu-

nications system manufacturer. Meanwhile their national manufacturers were not at all successful in exporting their products, and were confined to their domestic markets.

The German and French governments perceived an opportunity for their national manufacturers to take a dominant position in the upcoming 2G mobile communications market. They aimed to subsidize their national companies to develop a 2G technology and to push it to be adopted as the pan-European GSM technological specifications.

It is important to note that both the French and German governments interpreted the market needs and opportunities to be concentrated in urban areas. This bias was perhaps motivated by the fact that, as sketched in their historical backgrounds, their communications infrastructures had been built as an urban-centric structure.

Thus, their interpretations were driven by (1) their ambition to promote and protect their electronics industries, more specifically, mobile communications system manufacturers, and (2) their perceived need to maintain and leverage their existing communications structure. These interpretations were eminently reasonable, considering their markets and political power and the capabilities of their national manufacturers. The question was whether they were able to formulate the right standards strategies.

2.3.4. West German and French Governments' Strategies

Configuration of Value Network

Based on their situations and their interpretations, the German and French government subsidized the formation of four different consortia composed of various German and French companies to develop candidate technologies for the GSM standard (see Table 3).

With regard to developing the technology, the German and French governments basically outsourced the development to their national electronics firms, thus giving them opportunities to gain and improve their capabilities and resources in developing and manufacturing mobile communications systems.

In accordance with the urban-centric interpretations of the German and French governments, these German-French consortia developed and proposed technologies and systems mostly tailored to densely populated urban areas like Frankfurt and Paris.

To ensure that the GSM standard, which they expected to control, would indeed become a pan-European standard, the German and French governments persuaded Italy and UK, which had large markets as well, to also adopt the same technology that would be approved by the GSM group.

Formation of Standard Setting

Setting a common standard through the GSM group would be acceptable to all actors in Europe because of the pan-European character of CEPT. Given this fact, the correct strategy of the German and French governments was to standardize their technology through the GSM group. Working in the GSM venue, they advocated their technologies to be selected for the specifications of the standard. To garner the support of the voting members of the GSM group, the German and French governments also ought to have made sure that the technologies they proposed in fact met the interests of the others, particularly the Nordic PTTs.

Openness of IPR (Intellectual Property Rights)

The German and French governments did not have explicit policies for IPR on technologies developed by the consortia they supported. The companies participating in the consortia therefore had the opportunity to gain IPR on certain technologies for their own benefit. These companies, who were used to enjoying special arrangements with PTTs in the traditional communications industry, naturally continued their conventional IPR strategy which was to develop closed proprietary technologies

suitable for vertically integrated mobile communications systems. To them, it was unattractive to develop open interfaces that would facilitate interoperability between systems, because open interfaces and open IPR would allow new entrants to compete with them and cut their profits.

Although the German and French governments did not directly get involved in IPR strategy, their inaction basically allow the closed IPR strategy of their subsidized manufacturers to prevail. Consequently, the standards that the German and French governments proposed did not support the open horizontal value network vision of the GSM group.

2.4. Ericsson

The Swedish firm Ericsson was a major actor in setting the European 2G mobile standard. Indeed, the GSM specification adopted in 1987 was essentially an outgrowth of Ericsson's technology proposal. This section describes the standards strategy of Ericsson, which enabled it to prevail over its seemingly more powerful rivals.

2.4.1. Background of Ericsson

Ericsson was founded in 1876 as a telegraph equipment repair store in Sweden. Ericsson grew gradually and became a great success in manufacturing manual telephone exchanges. By the late 1890s, the small Swedish communications equipment market became saturated, prompting Ericsson to enter foreign markets including those of Britain, the U.S., and other Nordic countries. During its international expansion, Ericsson established factories in other nations and built systems according to the host nation's technology standard. Consequently, Ericsson became one of the first major global communications equipment providers in the world. This brief historical background showed that Ericsson as a company was oriented to working with other nations in order to grow its business.

The following analysis of the situation and interpretation of Ericsson is concerned with the time frame 1982-1987 in which a pan-European 2G standard was to be developed and adopted by the GSM group.

2.4.2. Ericsson's Situation

Organization's Capabilities to Meet Market Needs and Opportunities

To compete beyond the small market of Sweden, Ericsson had developed the capabilities to export communications systems to other nations. Based on its experiences in standardizing and implementing NMT, Ericsson had been manufacturing not only NMT-based systems but also systems based on other 1G technology standards.

The Swedish PTT was one of the PTTs that suggested CEPT to develop a pan-European mobile communications technology. As a Swedish manufacturer, Ericsson had worked intimately with the Swedish PTT. Therefore, Ericsson was in a position to comprehend the strategic direction of the GSM group within CEPT.

All in all, Ericsson's experience, technical competence, and strategic and market awareness gave it the capacity to explore the market opportunity of an open pan-European mobile technology.

The Position of an Organization in the Market

Sweden was a small country and was not a member of the EU at the time. Neither Ericsson nor the Swedish PTT was in a strong position to pressure other actors. However, Ericsson was widely acknowledged as a leading mobile communications system supplier and capable of supplying major PTTs irrespective of the telephony standard. In the mobile sector, Ericsson had been supplying NMT systems across the Nordic region, AMPS systems in the U.S., and TACS (Total Access Communication System – a variant of AMPS) in the U.K.

Thus, Ericsson was in the unique position of having good relations with PTTs from many countries, and therefore was in position to advise and influence the many PTTs in their choice and support of the 2G standard.

Ownership and Deployment of IP (Intellectual Property)

As an early participant in developing a proposal for the GSM standard, Ericsson had patents on various technologies. Some of them could have become essential patents if its proposal was selected to become the technological specification of GSM. However, when ETSI requested companies to claim essential IPR on GSM in 1996, Ericsson did not claim its IPR. According to the ETSI IPR Database, Ericsson's declaration of its essential IPR on GSM began to show in January 2001 (ETSI, 2008).

Ericsson had the same situation as the GSM group and the West German and French governments for the following elements:

- The availability of complementary products or compatibility of products/services.
- The type of technology innovation.
- The availability of alternative or substitutable technologies.

2.4.3. Ericsson's Interpretation of its Situation

Because of its historical background, Ericsson perceived very early on the potential of a huge market if CEPT adopted an open European 2G standard. The fact that most national manufacturers were opposed to the idea of having a single mobile communications standard in Europe could work to Ericsson's advantage, if Ericsson could convince the PTTs that an open pan-European standard would actually benefit them.

2.4.4. Ericsson's Strategies

Configuration of Value Network

Since 1977, Ericsson, Televerket (the Swedish PTT) and several Swedish universities had researched various possible technologies for the 2G standard. As a result, Ericsson knew the pros and cons of various technologies, particularly the difference between broadband and narrowband TDMA air interfaces (Bekkers, 2001; Meurling and Jeans, 1994).

The technology Ericsson deliberately chose was the narrowband TDMA. Overall, the narrowband TDMA system was most suitable for providing medium capacity in average or sparsely populated regions (Bekkers, 2001). In particular, the narrowband design was much more cost-effective than the broadband design with respect to building out the network (e.g. fewer base-stations would be needed). Since most of the European providers operated in areas that were not densely populated, and since the cost of implementation was the one of the most significant issues for service providers in adopting a technology, Ericsson's technology matched well with the needs of most of Europe's service providers. One prong of Ericsson's standards strategies was to lobby hard for the support of these PTTs.

Ericsson was keenly aware of the great power of the German and French governments to pressure their national PTTs to side with the technology the two governments were subsidizing. Thus, a pan-European technology standardization would be impossible without concessions from the German and French governments, even if the GSM group did select the technology that Ericsson proposed.

Expecting opposition and resistance from the German and French governments, Ericsson made certain preemptive moves to undermine the position of the German and French governments. It first agreed with Siemens (a major German national manufacturer) to cooperate in developing digital telephony technologies to implement whatever pan-European standards that the GSM group might adopt. It also agreed with LTC (a French company participating in one of the German-French consortia) about exchanging technologies. This technology exchange actually involved the French PTT, that later chose to support the narrowband technology (Bekkers, 2001). By getting German and French involvement in the narrowband TDMA technology, Ericsson weakened the rationale of the German and French governments' opposition as based on national interests.

One may say that Ericsson's strategy in shaping a value network was to reach out to those other actors, both operators and manufacturers, that would benefit from Ericsson's proposal, and whose collective support would provide a critical mass for success.

Formation of Standard Setting

Ericsson's strategy was to promote its version of a pan-European standard through the GSM group. Superficially, Ericsson's standards strategy was the same as the German and French governments regarding the format of standard setting. However, while Ericsson aimed to steer PTT supporters to assert their voting rights in choosing the GSM standard, the German and French governments planned on using their size and influence to overawe the PTTs.

Openness of IPR (Intellectual Property Rights)

Consistent with its strategy in shaping the future value network, Ericsson was willing to share and exchange technologies with others, so as to make its proposal more attractive. Thus, Ericsson's IPR strategy was to promote open IPR and RAND licensing. This was diametrically opposed to the strategy of the manufacturers from the consortia that the French and German governments subsidized.

2.5. Comparison of Actors in the GSM Standard

The following three sections compare the standards strategies of the major GSM actors, and discuss how the interaction of their strategies led rise to the final agreement on the GSM standard.

2.5.1. Comparison of Actors' Situations (GSM)

The situation of the GSM group, that of the German and French governments, and that of Ericsson were largely similar in all elements except the actors' market positions. Specifically, the technology-related elements (viz., the availability of complementary products or compatibility of products/services, the type of technology innovation, the availability of alternative or substitutable technologies, and ownership and deployment of IP) were similar to all the actors. After all, they referred to the same technology environment of

developing a 2G standard in the 1980s. Aiming in their own ways, all the actors recognized and had the capabilities to meet market needs and opportunities for the 2G mobile communications technology. However, because they had very different interests and market positions, their interpretations of the situations were based on very different aspirations, and hence their strategies were different (see Table 1).

2.5.2. Difference in Position and Power (GSM)

The GSM group as one of the working groups in CEPT had the statutory power to select a 2G mobile technology and set the technical standard specifications on behalf of the European PTTs. The power and interests of the GSM group were ultimately derived from its PTT sponsors. As a collective, the PTTs had the power to go against individual governments.

Table 1. Comparison of actors' situations (GSM)

Element \ Actor	GSM group	French and West German governments	Ericsson
Organization's capabilities to meet market needs and opportunities	(S) Along with political and economic unification in West Europe, all actors recognized the market needs and opportunities in developing a pan-European technology standard. They had capabilities directly or indirectly to meet these needs and opportunities.		
The availability of complementary products or compatibility of products/services	(S) The actors needed to develop not only complementary but also core systems for the GSM standard.		
The type of technology innovation	(S) The GSM technology was disruptive innovation based on digital telephony, different from the existing analog technologies.		
Market position	(D) The GSM group consisting of monopoly PTTs in Europe could leverage great power if all members agreed and committed to developing or adopting the same technology.	(D) Germany and France were two of the most populated countries in West Europe. With their market size and the ownership of their monopoly PTTs, the two governments had position to control their regions.	(D) Although Ericsson had a reputation as a manufacturer in the world, it could not force any nation to adopt its technology.
The availability of alternative or substitutable technologies	(S) The NMT-900 had been adopted in some countries. Some people considered it an alternative to GSM, because it used the same frequency bands – 900 MHz.		
Ownership and deployment of IP (Intellectual Property)	(S) Decisions on the technological specifications for the standard would affect the IPR ownership of essential and non-essential technologies. However, it was too early to identify which actors owned the essential IPR before the GSM group selected a proposal and finalized the technological specifications.		

[(S) Similarity, (D) Difference]

The German and French governments had the power to control their own large domestic markets. Without these governments' consent, it would not be possible to realize a pan-European mobile communications technology standard. Regardless of the GSM group's desire to standardize one technology, the European mobile market would still be fragmented if the German and French governments developed and implemented a different technology for their markets.

Ericsson had the least power among these actors. It did not have the position or power to pressure any service providers or governments to adopt its technology. To gain influence, it had to orchestrate and leverage the interests and power of the other actors.

2.5.3. Comparison of Actors' Interpretations and Aspirations (GSM)

In contrast to the similar situations that the actors confronted, their interpretations of the situation were very different.

The GSM group interpreted its situation from the perspective of a user or adopter of technology. Representing the interest of the PTTs, the GSM group's main concerns were costs and market opportunities. The nature of the technology itself was not of high interest to the group. The group's aspiration was to agree on and adopt an open pan-European mobile communications technology standard that (1) could provide international calls by interconnecting national networks, (2) that could be implemented cost-effectively while gaining economies of scale to reduce costs, and (3) that would not lock the PTTs into the proprietary technologies of a few manufacturers.

The German and French governments interpreted their situations from the point of view of being a powerful actor that could influence the development and implementation of a technology in two of the largest markets in Western Europe. Their concern was to boost their electronics indus-

tries. Their aspiration was to have the technology developed by one of the consortia they subsidized be adopted as the pan-European standard. If that came to be, then their national communications system manufacturers would gain competitive advantage and the opportunity to dominate the mobile technology market.

Meanwhile Ericsson interpreted its situation as a manufacturer that did not have much power to pressure other actors. Therefore, it had to be pragmatic and opportunistic. Ericsson's main concern was to align itself with the interests of other actors in terms of their needs and aspirations in standardizing the 2G technology, so that the other actors would support Ericsson's proposed technology. Ericsson's aspiration was to rally the PTTs to oppose the German and French governments and adopt an open standard.

These major actors interpreted their similar situations differently based on their own values, concerns and aspirations. These interpretations shaped their strategies.

2.5.4. Comparison of Actors' Strategic Directions (GSM)

The following highlights succinctly the strategic direction of the major actors (see also Table 2).

1. The PTTs via the GSM group strived to configure an open horizontal value network for 2G mobile service. The GSM group pushed for a pan-European standard that was agreeable to the PTTs.
2. The dominant national manufacturers passively dragged their feet about the pan-European technology standardization. They would rather not compete against others under an open standard.
3. The German and French governments encouraged their national manufacturers to develop a technology for the GSM standard.

Table 2. Comparison of actors' strategic directions (GSM)

Actor / Aspect	GSM group	German and French governments	Ericsson
Configuration of the value network	(D) The GSM group's strategy was to create a horizontal value network based on the open interface standard.	(D) These governments' strategy was to create German and French manufacturer-centered consortia to develop a technology based on the wideband design. They also configured the political network to create a larger market by agreeing with Italian and British governments.	(D) Ericsson's strategy was to consider other actors' needs and reach out them to be a part of the network it formed.
Formation of standard-setting	(S) All actors wanted to standardize the technology through the standard-setting organization, the GSM group.		
Openness of IPR	(D) The GSM group preferred open interfaces between systems with open IPR to remove lock-in effects from manufacturers.	(D) The manufacturers involved in the consortia want to control their IPR tightly on the technologies they developed.	(D) Ericsson was willing to share its technology with other actors to promote the technology, so its IPR strategy was rather open.

[(S) Similarity, (D) Difference]

4. Instead of reaching out to as many actors as possible, the German and French governments formed and subsidized national consortia to develop technology that focused on providing mobile service in densely populated urban areas.

5. The ownership of IPR on technologies developed by the consortia went to companies that participated in the consortia.

6. The German and French governments obtained agreement with the governments of Italy and UK to support the technology the GSM group would adopt.

7. Ericsson aimed to provide the technology that met the PTTs' needs. Ericsson promoted the narrowband design, TDMA, which was cost-effective for mid- or less-populated regions, thereby gaining the crucial support of the PTTs.

8. Ericsson was prepared to share IPR and formed partnerships with other firms. Through this strategy, Ericsson gained German and French allies. Moreover, Ericsson was in effect realizing the open horizontal value network that the GSM group desired.

2.6. Emergence of the GSM Standard and the European 2G Configuration through Actors' Interactions

At the level of industrial structure, the strategy of each actor entails the pursuit of a certain configuration of the future value network – which benefits some and disadvantages others. The actors' strategies interact in the sense that they each choose to accept, reject, adapt to, compromise with, or transform the proposed value network configuration according to how their own aspirations and benefit might be realized.

A configuration of the industry emerges when enough actors accept a certain role in the value network. (The term "self-organized" is appropriate when the configuration comes about without central direction.) Obviously, the strategy of an actor is more likely to be successful if the strategic direction of the development of the industrial configuration is attractive to many other actors, and if the value-network the actor proposes is both robust and flexible in accommodating others' interests.

In the early 1980s, the major actors of the European 2G standardization (GSM group, the French and German governments, and Ericsson) strived to formulate and pursue what each regarded as the right standards strategy. They each explicitly or implicitly advocated a certain configuration of the future value network of the 2G mobile communications. When the GSM group invited all interested actors to propose by the end of 1986 a demonstration system and technology for consideration to be adopted as the pan-European standard, the actors had to show (through their choice of technologies and partners) in no uncertain terms what their visions of the future 2G value network were.

The GSM group, as a representative of the monopoly PTTs in Europe, was in the position of a powerful buyer of technology. And as the standard-setting organization for the pan-European mobile communications technology, the GSM group had the obligation to open its consideration to all European proposals. The group requested the interested actors to submit their demonstration systems by the end of 1987, so it could evaluate them and select the most appropriate technology. The PTTs were in principle neutral on who might turn out to be the technology suppliers in the future 2G value network. They were mainly concerned with shaping a value network configuration in which the PTTs could increase revenues, decrease costs, and not be locked into manufacturers.

Eight demonstration systems were proposed – two from non-manufacturers (the Swedish operator Televerket (now Telia), and Elab research laboratory from Norway) and the rest from manufacturers. As listed in Table 3, the first four proposals were from the various consortia subsidized by the German and French governments, and the last four proposals came from the Nordic countries.

The GSM group evaluated the submitted demonstration systems in the facilities of CNET – the research institute derived from the French PTT (Noam, 1992; Bekkers, 2001). Meanwhile, the PTTs also conducted their own tests to evaluate feasibility, estimate costs, and validate the service quality and frequency efficiency of the proposals. Based on extensive evaluations, two of the eight proposals were chosen for further consideration – the SEL/Alcatel proposal (subsidized by the German and French governments) and the narrowband TDMA proposal from Ericsson.

These two proposals, if adopted, entailed very different configurations of the 2G value network.

Table 3. Proposals for the GSM standard

German-French consortia	
Developer	**Technology**
"SEL/Alcatel" – SEL, AEG/Telefunken, ATR, SAT, Italtel	Wideband TDMA / CMDA intracel separation and frequency reuse
"Philips" – TeKaDe / TRT	Hybrid TDMA/FDMA/CDMA
ANT/Bosch/Matra	Narrowband FDMA/TDMA
LCT/TRT	TDMA with slow hopping CDMA and FDMA intracel
From the Nordic region	
Developer	**Technology**
Ericsson	Narrowband TDMA
Televerket (now Telia)	Narrowband TDMA
Nokia	Narrowband TDMA
Elab	Narrowband TDMA
Note: Although some developers used the same technology like TDMA, the characteristics of their proposal were different, for example, the various number of allocated time slots.	

(Source: Bekkers, 2001, p. 288)

It is useful to describe and contrast these configurations.

The SEL/Alcatel proposal:

1. The SEL/Alcatel proposal was designed for densely populated markets. It suited the German and French markets and fitted with current communications infrastructure, which historically had focused on serving metropolitan areas.
2. Many European cities and regions were medium or sparsely populated. SEL/Alcatel's proposed technology was not cost-efficient for service providers that mainly operated in these areas.
3. SEL/Alcatel's proposed technology was essentially proprietary (Bekkers, 2001; Cattaneo, 1994; Iversen, 1999). The German manufacturer SEL held significant patents on technology essential to the proposal – for instance, EP patent 257110A1 filed in 1986 (Bekkers, 2001).
4. The proprietary nature of the SEL/Alcatel proposal raised barriers to entry for the other suppliers, as well as exposing PTTs to the risk of lock-in effects.

The Ericsson narrowband TDMA proposal:

1. Ericsson's proposed narrowband technology was more cost-effective for many European service providers than SEL/Alcaltel's technology.
2. It was attractive even to the French and German PTTs, because the French and German PTTs needed to build new GSM infrastructure for their entire nation which included many sparsely or averagely populated areas.
3. When Ericsson's proposal became one of the two finalists, Ericsson declared its intention to open up its IPR, and formed partnerships with Siemens and LTC.

4. With open IPR, Ericsson's proposal was more attractive to other manufacturers. When many manufacturers could compete for business, an open horizontal value network would be realized.

When the future 2G value network as entailed from the two proposals were compared, it was clear that the GSM group, if faithful to its character, would select Ericsson's proposal. Indeed, the GSM group's preference for narrowband TDMA became clear by 1987. But whether this technology and the value network it entailed would be realized as a pan-European standard depended on whether the German and French governments would concede. This turned out to be a matter of EU politics.

2.7. Final Realization of the 2G GSM Configuration

Considering the two alternative configurations of 2G value networks, many members of the GSM group much preferred Ericsson's narrowband-based proposal over SEL/Alcatel's wideband-based proposal (Bekkers, 2001). However, the German and French governments were strongly against this preference, even though their own PTTs favored Ericsson's proposal (Bekkers, 2001; Cattaneo, 1994). Considering the traditionally tight relationship between a government and its PTT at the time, it was remarkable that the German and French PTTs expressed and even acted against their governments' wishes.

To break this deadlock situation, the German PTT, Deutsche Bundespost, organized a "hearing" to assess the advantages and disadvantages of the two proposals. The result of the hearing again favored Ericsson's narrowband design. The support of German Siemens and French LTC for Ericsson's proposal further weakened the stand of the German and French governments.

Finally, the GSM group fixed the outline of the GSM standard. This final standard provided broader technical specifications than any one proposal suggested, thus gaining broader support for the GSM standard. In this final outline standard, it was obvious that the wideband design was rejected, as the standard was essentially based on the Ericsson and Televerket proposals (Bekkers, 2001). This decision was difficult for the German and French governments to accept, but this GSM standard was acceptable to the British and Italian governments.

However, the lack of support from the French and German governments precluded the standardization of this technology as the pan-European mobile communications technology. Therefore, persuading the German and French governments to support the GSM standard was in the interest of all other actors. Once the GSM group approved the technological specifications, it was time to politically persuade the German and French governments to agree.

The British government, that had once supported the strategy of the German and French governments, initiated two diplomatic efforts to resolve this disagreement: 1) the British government arranged a meeting with the delegates of the French and Italian governments; and 2) Geoffrey Pattie, the U.K. Minister of Trade and Industry had a meeting with Christian Schwarz-Schilling, the German Telecommunications Minister (Bekkers, 2001; Garrard, 1998). What exactly transpired in the meeting between the two ministers was not publicly known. But it was reasonable to surmise that Germany and France were persuaded to look beyond their proposed 2G value network and consider the large benefit to the EU, whose unity and prosperity Germany and France had been championing.

Eventually, after the ministers of four countries (Germany, France, U.K. and Italy) met together on May 19, 1987, these countries announced that they would support the decision of GSM group with the minor (face-saving?) request to change small technological specifications related to the modulation technique (Bekkers, 2001).

2.8. Lessons from GSM 2G Standardization

This analysis of organizations' standards strategies in the GSM standardization process and the industrial self-organized evolution illustrates the viability and necessity of the proposed Self-organized Complexity Unfolding Model and Framework of Organizational Standards Strategy:

- First, actors had their situations affected by the 1G industrial configuration. The Unfolding Model and Framework of Organizational Standards Strategy helped to understand the nature of the 1G industrial configuration and how it affected each actor's situation.

- Second, they interpreted their situations and envisioned a future 2G mobile industry and market. Based on these interpretations and visions, the actors developed standards strategies to form a competitive value network. Using the Framework of Organizational Standards Strategy, we were able to analyze actors' situations, interpretation of the situations, and standards strategies.

- Third, the interactions of these value networks that the actors tried to form based on their standards strategies influenced the self-organized configuration of GSM standardization. Again, applying the Self-organized Complexity Unfolding Model, it was observed how the self-organized configuration emerged.

As shown the case of GSM standardization, organizational standards strategies are very complex, organic and dynamic. It is impossible to analyze this complex phenomenon by applying simple equations. An applicable model or framework should be flexible enough to consider complex and dynamic environments, as the proposed model and framework do.

GSM standardization is a very complex phenomenon to analyze. However, it was only one part of the 2G mobile communication industry. Another part came from the standardization of IS-95 (cdmaOne). The same method will be applied to analyze IS-95 standardization. After that, we can finally move to a higher level to analyze the overall industrial configuration of 2G standards by applying the Self-organized Complexity Unfolding Model.

3. THE NEW AMERICAN 2G STANDARD: IS-95 (CDMA)

In the U.S., two 2G mobile communications standards with completely different air interface technologies were adopted and commercialized by operators. The D-AMPS standard, adopted by AT&T and others, was an evolution of the 1G AMPS technology based on TDMA air-interface. The IS-95 standard, adopted by PacTel and others, was a revolutionary technology based on the CDMA air-interface (IS-95 [i.e. Interim Standard 95] is the official name of this first ever CDMA-based cellular telephony standard. It is also known as "TIA-EIA-95" and its trade name is "cdmaOne").

In this book, based on three reasons, the IS-95 standard will be focused. First, the development and introduction of IS-95 illustrates the kind of aggressive standards strategy required for a new technology to compete against an entrenched technology and standard. Second, the CDMA technology later on became the foundational technology for the next generation (3G) mobile standard. To prepare for our subsequent study of 3G standardization, we must understand how the IS-95 standard got established. Third, the IS-95 value network was more global and intricate than that of GSM or D-AMPS and was the precursor of a very complex 3G value network.

3.1. Background of the IS-95 Standard

With the breakup of AT&T's monopoly and the U.S. government's policy of granting two licenses in each region, which consequently allowed many mobile service providers to enter into the U.S. market, the U.S. mobile market became much more open and competitive. Since AMPS was the only 1G mobile communications technology available, the competing providers did not have to worry about setting standards. However, the openness of the market and the U.S. government policies favoring open competition turned out to have a major effect on the standardization of 2G mobile communications technology in the U.S. In theory, mobile service providers were free to choose any 2G technology (Note that, in sharp contrast, all European service providers were mandated to adopt GSM). This freedom to choose technology was the pre-condition for the introduction of the revolutionary CDMA technology to the U.S. mobile market.

In the U.S. market, the number of subscribers for 1G mobile communications service had grown rapidly ever since AT&T Illinois Bell commercialized the AMPS technology in 1983. Many actors, especially service providers like PacTel (now Verizon) that operated in urban areas, soon recognized the need for a 2G mobile technology that would provide more capacity than the 1G AMPS technology.

Manufacturers such as AT&T and Ericsson recognized the limits of AMPS and conducted research to find a suitable 2G digital technology throughout the 1980s. Their technology strategy was to evolve a 2G technology from the 1G technology. This was at the time a reasonable strategy because 1) a new technology that fit easily with the existing network would minimize development and implementation costs; 2) operators needed to support the existing customers of the 1G mobile systems; 3) no new spectrum was forthcoming, so 1G and 2G had to share the same frequencies (Mock, 2005); and 4) most significantly, the manufacturers that had developed the 1G technologies were more inclined to improve their existing technologies with their existing resources and capabilities rather than investing in totally new resources and capabilities necessary for developing a disruptive innovation such as CDMA.

Thus, these manufacturers developed the D-AMPS (Digital AMPS). The D-AMPS technology could increase the capacity of AMPS infrastructure by splitting the frequency channel of a single user to allocate it to three or more users by time slots. Thus, the D-AMPS technology was a continuous innovation based on AMPS. Although the D-AMPS technology could offer three times more capacity than AMPS, having more than three users share the same channel could greatly stress the networks, adversely affecting connectivity and the quality of calls. Therefore, beyond the three-fold increase in capacity, D-AMPS as a technology did not have any "head-room" for further improvements.

Although the D-AMPS and the GSM technologies were based on the TDMA technology, their specifications were different and they were incompatible with each other. The American actors, used to being global leaders in the communications industry, did not even consider adopting the European standard of GSM even though their development of 2G technology occurred later than the European actors when GSM was available.

The CTIA (Cellular Telecommunications and Internet Association), which was founded in 1984 as an international not-for-profit organization to represent the interests of communications companies, investigated the market need for 2G mobile communications technology and encouraged the industry to develop a technology that could meet the User Performance Requirements (UPRs) that CTIA established (Mock, 2005). The UPRs included "1) a capacity of at least 10

times greater than the existing AMPS standard; 2) support for a gradual transition from existing AMPS networks; 3) low cost penalties; and 4) rapid implementation" (Bekkers, 2001, p. 349).

The CTIA was not a formal standard-setting organization but its support for a particular technology could influence the actors' perceptions and their technology and standards strategies. This was also true for the U.S. voluntary standard-setting organization TIA (Telecommunications Industry Association).

Although the D-AMPS technology could not meet CTIA's capacity recommendations for 2G mobile communications technology, many actors supported it as a feasible technology to develop and implement. For this reason, TIA voted for D-AMPS (TDMA method) for the 2G mobile communications technology standard in the U.S. market. The D-AMPS standard was formally endorsed by CTIA as well in January 1989 (Mock, 2005).

Thus, by the late 1980s, it was widely accepted that D-AMPS would be the 2G mobile communications technology standard in the U.S. At this late stage of the U.S. 2G standardization process, the small technology company Qualcomm proclaimed that it had a much better technology for 2G. It began to advocate its CDMA technology at various conferences and meetings in 1988 (Mock, 2005) right after CTIA released its set of UPRs. Although the CDMA technology seemed technologically superior and promised to fulfill CTIA's UPRs better than D-AMPS technology, it was totally new and unproven. Since the operators and manufacturers had already invested in the development of D-AMPS, at the outset they were not inclined to consider Qualcomm's CDMA as a 2G technology standard.

The CTIA and TIA endorsements of D-AMPS and the initial cold reception of CDMA should have discouraged Qualcomm. However, Qualcomm did not give up the commercial possibility of CDMA. Because the decision of TIA was not compulsory, and because the U.S. government agent, the FCC, did not mandate D-AMPS, Qualcomm was free to directly approach service providers to persuade them to adopt CDMA. In fact, Qualcomm's aggressiveness and doggedness in the end paid off handsomely. The section of QUALCOMM (IS-95) will introduce Qualcomm's strategies based on its situation and interpretation of how best to push the CDMA technology as a 2G standard given its unfavorable starting position.

Without the help of mobile service providers, Qualcomm would not have been able to push its CDMA technology because it did not have enough resources and influence. The U.S. mobile service providers that chose to help Qualcomm by and large had similar situations, interpretations and strategies. They were usually based in large urban areas – PacTel in Los Angeles, NYNEX Mobile in New York City, and Ameritech in Chicago – and they required much larger capacity than D-AMPS could likely provide. Among these mobile service providers, PacTel was the first mobile service provider that invested in Qualcomm's CDMA development and had a field test with Qualcomm. Thus, as a representative of these mobile service providers, PacTel's situation, interpretation and standards strategy will be presented in the section of PACTEL (Pacific Telesis Group).

Organizational strategies concerning the D-AMPS standard will be not included in this book. For the actors that evolutionarily accepted the D-AMPS technology as a natural migration from 1G AMPS, their focus was to develop and implement the 2G D-AMPS technology and not to determine standards strategy (which is the subject of this book). However, their great resistance against the standardization of the CDMA technology will be mentioned in subsequent sections.

Outside the U.S., a very important actor in establishing CDMA as a 2G technology standard was the Korean government. In the section of The Korean Government (IS-95), the background, situation, interpretation, and strategy of the Korean government with respect to mobile communications will be described in detail. Here, it

would be only pointed out that Koreans made a big bet on CDMA technology. The bet and their efforts began to pay off with the hugely successful commercialization of 2G CDMA mobile services in 1996. The Korean success proved the stability and great capacity that the CDMA-based technology could offer. Witnessing this concrete evidence, many actors started to lift their doubts and confidently join the value network created by Qualcomm and the Korean government. This value network later provided the platform from which Qualcomm and the Korean actors strived to influence and shape 3G standards.

Lastly, it should be noted that Qualcomm developed the IS-95 CDMA technology all by itself and therefore gained exclusive patent rights on the technology. So unlike the case of GSM development, competition to secure and capture IPR value was not an issue in establishing the IS-95 standard. The involved actors were concerned rather with the pragmatic issues of commercializing and diffusing this CDMA technology which came on the scene much later than GSM and D-AMPS.

3.2. Qualcomm (IS-95)

3.2.1. Background of Qualcomm (IS-95)

In 1985, Qualcomm was found by seven people who had worked at Linkabit, which started as a consulting firm in communications technology. Qualcomm was mostly involved in government contracts related to military and space projects, and honed its innovation through these projects. In particular, consulting on a Hughes Aircraft proposal to bid on a license for a global satellite communications service helped Qualcomm to develop the technique of spread spectrum air interface based on CDMA technology. Through these kinds of innovations, Qualcomm started to accumulate patents on mobile communications applications using CDMA technology in 1986. Although the project with Hughes Aircraft delayed and died due to other external factors, Qualcomm saw the opportunity to apply the same techniques to terrestrial mobile communications. The mobile communications market was growing rapidly and began to reach the limited capacity of the AMPS technology in urban areas. But by the time Qualcomm started promoting the CDMA technology in 1988, it was already late (see Figure 2).

Many manufacturers such as AT&T and Ericsson had already invested in evolving the 1G AMPS technology to the D-AMPS 2G technology. These companies were set to push the D-AMPS technology as the U.S. 2G standard. Even though the D-AMPS was different from the GSM technology, both were derived from TDMA technology. Therefore, European manufacturers such as Ericsson and Nokia could readily utilize their capabilities and resources in GSM to develop and produce D-AMPS systems. With the support of the major manufacturers, D-AMPS was widely acknowledged in the industry as the de facto U.S. 2G standard. However, some operators such as PacTel concerned that D-AMPS could not provide all the capacity they needed in urban areas.

Qualcomm did not give up hope even after TIA selected D-AMPS as the 2G technology standard in 1989 for the U.S. market. But it had to face up to its adverse situation in 1989 and formulate a promising strategy.

3.2.2. Qualcomm's Situation (IS-95)

Organization's Capabilities to Meet Market Needs and Opportunities

As a new innovative entrant, Qualcomm had strong capabilities to develop a superior technology that would meet future market needs and opportunities. But its resources were too limited to develop and commercialize a CDMA-based application by itself.

The Availability of Complementary Products or Compatibility of Products/Services

The CDMA technology was unproven at the time. There were no complementary products. To deliver a working CDMA system, Qualcomm would have to build everything from scratch.

The Type of Technology Innovation

CDMA was a disruptive innovation in the mobile communications industry, not based on any previous technology.

The Position of an Organization in the Market

As a new, small, and unknown entrant to the mobile communications industry, Qualcomm did not have any power to influence others to adopt the technology it would develop.

Figure 2. The timeline of development of IS-95 standard

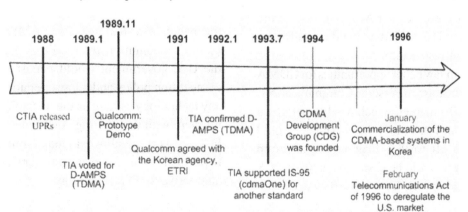

The Availability of Alternative or Substitutable Technologies

When Qualcomm started advocating for its CDMA technology in the late 1980s, other technologies – GSM, PDC and D-AMPS – were well on their way to deployment in Europe, Japan and the U.S. respectively. Therefore, alternatives to CDMA were already available, even though they were not yet commercially available.

Ownership and Deployment of IP (Intellectual Property)

Although the CDMA concept was invented by two musicians during World War II, Qualcomm filed and gained patents for the essential technologies necessary for commercial CDMA-based mobile applications. For example, Qualcomm acquired patents on the method of power control necessary for CDMA signal reception, and it also held a patent on the soft handover of calls between cells. Both patents were critical to making CDMA telephony work (Mock, 2005).

3.2.3. Qualcomm's Interpretation of its Situation (IS-95)

As a technology innovator, Qualcomm believed that CDMA technology was a better 2G communications standard that could meet all market needs and opportunities. Although Qualcomm believed CDMA would work well in the existing 1G spectrum, it realized that the 1G operators would rather preserve their AMPS infrastructure and migrate to D-AMPS than build a totally new 2G CDMA infrastructure.

Qualcomm saw better opportunities for CDMA in the yet-to-be formed Personal Communications Services (PCS) market, which would operate on a high frequency that the U.S. government would soon auction off to PCS licensees. Qualcomm believed that it would be easier to persuade PCS operators to adopt CDMA because they would not

have the burden of path dependence – they did not have any existing infrastructure to preserve.

Qualcomm perceived another opportunity in Wireless Local Loop (WLL) systems for developing countries like China and India. Because of the high cost of laying landlines, the landline telephone systems in these countries were far from covering the countries well – urban areas were underserved and remote regions were hardly connected. WLL was a viable solution to connect these countries at a much lower cost. Qualcomm interpreted CDMA as a technology for WLL that could offer the additional capacity to deliver data as well as high quality voice service within a limited spectrum in these countries (Mock, 2005).

All in all, Qualcomm saw CDMA as a great technology that would serve markets better than all other technologies. But as a small and unknown company with an unproven technology, Qualcomm did not have the wherewithal to crack open markets. Therefore it had to persuade other more influential actors to risk giving CDMA a chance, and find ways to prove the viability if not the superiority of CDMA.

3.2.4. Qualcomm's Strategies (IS-95)

Configuration of Value Network

The mobile communications industry in the U.S. respected TIA's decision on D-AMPS (TDMA) and was prepared to adopt the 2G standard based on D-AMPS technology. Qualcomm had to make sure that there could be multiple technologies in the market instead of just one D-AMPS standard. First, Qualcomm visited the FCC to ensure that the U.S. government would persist in its open competition policy and allow operators to deploy any technology as long as the technology did not interfere with the existing AMPS networks. The FCC's assurance allowed Qualcomm to at least to persuade other actors to consider CDMA as an alternative for 2G technology.

Second, Qualcomm started to search for operators who were not satisfied with the capacity limitations of the D-AMPS technology, and would thus be more inclined to consider the advantages of adopting CDMA technology for 2G. PacTel (Pacific Telesis, which later became Airtouch Cellular and is now part of Verizon Wireless) showed interest in the CDMA technology when approached by Qualcomm. At the time, Pactel's chief scientist, Dr. William C. Y. Lee, was familiar with CDMA and was supportive of further developing the technology. He funded Qualcomm to build a prototype CDMA system. In November 1989, Qualcomm was able to demonstrate its prototype successfully and showed the practical possibilities of CDMA. After the demonstration, CDMA was no longer just a theory on paper.

Third, Qualcomm approached Motorola and AT&T to participate in developing the CDMA value network. Both companies were at the time heavily involved in developing the D-AMPS (TDMA)-based systems, so they requested tough conditions such as exclusive rights and a guarantee of firm orders for equipment business before they would license the CDMA technology from Qualcomm (Mock, 2005). Although their participation was lukewarm, they did sign licensing agreements with Qualcomm. Qualcomm also established relationships with Japanese manufacturers such as Kyocera.

Qualcomm's strategy in configuring a value network was (1) ensuring that the government would not enforce a single standard; (2) convincing operators who needed more capacity to consider CDMA; and (3) aligning with other manufacturers to help produce CDMA-based systems.

Formation of Standard Setting

After TIA voted for the D-AMPS (TDMA) technology as a 2G standard, Qualcomm created a two-prong strategy for establishing CDMA as an alternative standard. One prong was to pursue the de-facto path by persuading enough operators to adopt CDMA technology. Second, Qualcomm pres-

sured TIA to evaluate CDMA technology again as a standard by fulfilling the necessary CTIA requirements. Although Qualcomm was making a progress in persuading operators, support from TIA would help Qualcomm to persuade more actors (operators and manufacturers) to join its value network.

Openness of IPR (Intellectual Property Rights)

Although Qualcomm held multiple IPR for the essential technologies for CDMA systems, as a new small entrant, it did not have enough resources to develop and manufacture all systems to fulfill market demands. Thus, it decided to license its CDMA technology to manufacturers at a fairly low rate and assist them to produce mobile communications systems based on CDMA technology. By thus opening up its essential IPR on CDMA technology, Qualcomm aimed to build a value network that would assuage mobile service providers' concerns about supply shortages and supplier lock-in.

3.3. PacTel (Pacific Telesis Group)

3.3.1. Background of PacTel

PacTel was one of the seven communications companies created from the breakup of AT&T. It had the greatest market share in the West Coast and nearby regions through Pacific Bell and Nevada Bell. Pacific Bell was the biggest subsidiary among the twenty-two Bells in terms of assets before the breakup of AT&T.

PacTel provided mobile communications service in major cities like Los Angeles and San Francisco on the West Coast. When Qualcomm visited PacTel right after TIA's decision on D-AMPS, PacTel acknowledged that it needed new technology to support its growing market in urban areas. For example, the Los Angeles market was growing rapidly with over 100,000 subscribers. It was obvious that demand would soon overwhelm the capacity and call quality PacTel could offer (Mock, 2005). Even with D-AMPS technology,

PacTel predicted that it would not be able to provide sufficient capacity for the increase in subscribers in large cities.

However, PacTel knew that CDMA technology had not been proven with real-world applications, and many issues had to be resolved before CDMA could be commercialized. Thus, PacTel could not simply commit to implementing the CDMA technology. If PacTel committed itself to CDMA and the technology failed, then PacTel would go bankrupt.

3.3.2. PacTel's Situation

Organization's Capabilities to Meet Market Needs and Opportunities

As a major communications service provider, PacTel understood market needs and opportunities. PacTel worried that D-AMPS would not provide it with the capacity to provide service for the large number of subscribers in its fast growing markets. In addition to providing services to end-customers, PacTel also sold its own branded products under the PacTel name, even though it outsourced the manufacturing of telephones. It did not manufacture systems directly, but it could leverage manufacturers to provide the systems it needed even if they were CDMA-based.

The Position of an Organization in the Market

PacTel had a strong market position in the West Coast. Although it did not have the power to force other mobile service providers to adopt a certain technology standard, its selection of technology could influence other mobile service providers to choose the same technology.

Ownership and Deployment of IP (Intellectual Property)

PacTel did not have IPR on essential technologies for CDMA.

PacTel had the same situation as Qualcomm for the following elements:

- The availability of complementary products or compatibility of products/services.
- The type of technology innovation.
- The availability of alternative or substitutable technologies.

3.3.3. PacTel's Interpretation of its Situation

PacTel's interpretation of its situation was based on its pragmatic needs. It predicted that its existing systems and the approved D-AMPS technology by TIA would not be sufficient to meet the growing market in urban areas and that it might have to turn away new customers.

PacTel wanted a technology that could support its rapidly growing market. Given this situation, PacTel welcomed the possibility that CDMA technology could solve its capacity problem. But it had to make sure that CDMA would actually work.

3.3.4. PacTel's Strategies

Configuration of Value Network

Although PacTel perceived great promise in CDMA technology, it was risky for PacTel to commit to this technology when there were still unresolved technical issues (Mock, 2005).

After Qualcomm suggested solutions for the issues that PacTel pointed out, PacTel considered CDMA technology seriously. However, PacTel was still wary of committing. Thus, it started to invest financial and human resources to conduct a field test with Qualcomm and asked another systems manufacturer, Motorola, to cooperate in conducting the first field test. At the same time, PacTel also asked Qualcomm to persuade other mobile service providers to adopt CDMA, so it could avoid technological isolation (Mock, 2005). And it contracted independent consultants to evaluate the possibility of CDMA for future mobile communications.

PacTel's strategy in configuring a value network was to encourage Qualcomm to develop CDMA-based systems further and persuade other mobile service providers and manufacturers to support CDMA technology.

Formation of Standard Setting

It would have been very risky for PacTel to adopt CDMA technology if no other actors adopted it. If CDMA received endorsement from the standard-setting organization TIA, it would be easier to persuade other actors to adopt CDMA. Thus, PacTel encouraged and assisted Qualcomm in seeking approval from TIA.

Openness of IPR (Intellectual Property Rights)

PacTel was not sure that Qualcomm would have the capabilities and capacity to fulfill the demand for CDMA-based systems all by itself if the market grew quickly. In addition, as an adopter of the technology, PacTel did not want to depend on a single supplier. Thus, it preferred that there be multiple system suppliers. For this reason, PacTel asked Qualcomm to license its technology to other manufacturers (Mock, 2005).

3.4. The Korean Government (IS-95)

3.4.1. Background of the Korean Government (IS-95)

The Korean government has played a major role in the development of the Korean mobile industry since its beginning. To get a perspective of the Korean government's strategy on mobile communications, some background about the Korean government's involvement in industrial development in that nation will be introduced.

Korean industrial development first began during the Japanese occupation of Korea (1910-1945). It was concentrated in the northern part, where most of the peninsula's natural resources could be found. This meant that after division into

North and South Korea, South Korea had very little industrial development and few natural resources. The Korean War (1950–1953) left South Korea even more impoverished. This poverty and the lack of natural resources forced South Korea to try harder to develop economically.

During the Cold War period, Korea was not seen as an attractive country for investment by foreign investors. After Park Chung Hee gained power with a military coup in 1961, the South Korean government started to create economic development policies to solicit public support and cover its illegitimate power. The Korean market could not have free competition like the United States or western European countries, because there were not any Korean players that could compete with the giant foreign companies that had been in the market for a long time. From the government's perspective, the only way to develop the Korean economy was through a deliberate strategy to grow selected industries one by one in a protectionist manner.

The Korean government started by borrowing money from other countries and developing light industries for exports, like the garment industry. With the capital accumulated from the development of this sector, the Korean government was able to move into more sophisticated industries such as steel, ship building, and electronics. The companies that had close relationships with the Korean government were able to get into these industries and grow quickly under the government's wing.

By the 1980s, the South Korean market had become more open and the Korean companies had become more independent of government influence. Even though the Korean government could no longer intervene in industries and markets directly, it could still direct them by setting policies and regulations, and through the traditional tight relationships between government and industry leaders.

The first mobile communications company in Korea was founded by the Korean government in March 1984 as Korean Wireless Telecommunication. It started 1G mobile service based on AMPS technology in May 1984. Since then, subscribers to cellular-phone service have grown steadily, especially during the 1988 Seoul Olympics. However, the Korean electronics companies, manufacturers that had grown rapidly during the same time, barely contributed to and benefited from building the AMPS infrastructure and providing systems and devices. The goal of taking business away from foreign manufacturers such as Motorola and Ericsson, which dominated the 1G systems supply, appealed to both the Korean government and manufacturers.

When moving towards 2G technology, the Korean government had four choices for its future mobile industry – GSM, TDMA, PDC, or CDMA. The selection of GSM, TDMA, or PDC technology might have seemed natural, considering that all three technologies had been developed and were about to be implemented in other markets. However, these selections would have made the Korean mobile market continue to be dependent on foreign mobile manufacturers again as it was in the 1G period.

CDMA technology was developed by Qualcomm in 1989 but had not been introduced in any market. The U.S. and European markets appeared to have chosen to invest in D-AMPS (TDMA) and GSM technologies and rejected CDMA. However, if CDMA's theoretical superiority could be proven in a commercial market, it would give the first CDMA manufacturers and operators great advantages in the future development of the mobile industry. After Qualcomm successfully demonstrated the viability of CDMA to PacTel, the Korean government decided to bet on CDMA.

The Korean government figured that Korea could obtain significant technology transfer from Qualcomm, with low cost licensing, because Qualcomm was anxious to commercialize its CDMA technology. If this strategic bet on CDMA turned out to be successful, then the Korean electronics manufacturers would develop their competence by supplying the Korean mobile market, and be in position to compete worldwide.

In May 1991, the Electronics and Telecommunications Research Institute (ETRI) – founded by the Korean government in 1976 as a research institute – and Qualcomm agreed to jointly develop CDMA technology. This agreement implied that the Korean government and industry would invest enormous financial and other resources to develop all the necessary CDMA systems and commercialize CDMA technology. Betting on CDMA was an extremely risky decision for the Korean government. If the bet failed, the Korean government and Korean companies would end up losing a lot of time and investment, and they would have to build the mobile communications industry again using one of the other technologies (GSM, TDMA, or PDC).

Although the Koreans faced a lot of challenges in developing and implementing CDMA-based systems, through diligent efforts from 1991 to 1995, commercial CDMA mobile service was launched in 1996. After working out the bugs in the initial rollout of the CDMA network, the Korean industry found that CDMA technology was indeed capable of delivering robust and high capacity mobile service. From then on, the Koreans were on their way to become major actors in the global mobile industry.

The successful rollout of Korean CDMA service in Seoul might turn out to be one of the single most critical events shaping the world's mobile communications future. Although many people didn't and still don't realize it, it gave concrete proof to other actors that CDMA technology could provide more than enough capacity (with high voice quality) to one of the most densely populated cities in the world. Henceforth, the other actors would have to take CDMA technology and its leaders – Qualcomm and the Korean government and companies – seriously.

For analysis of the standards strategy of the Korean government, its situation and interpretation around the time of its agreement with Qualcomm in 1991 are presented below.

3.4.2. Korean Government's Situation (IS-95)

Organization's Capabilities to Meet Market Needs and Opportunities

With the steady growth in the Korean mobile communications market and the development of 2G technologies outside of Korea, it was obvious that the market was moving toward 2G technology. The question was which technology to adopt. The Korean government needed a technology that could not only satisfy the market growth but also boost its electronics industry. The Korean government also recognized that the migration from analog 1G to digital 2G was a critical opportunity for Korea to develop its competence in digital technologies.

The Position of an Organization in the Market

Although the Korean government did not have the capabilities to manufacture systems, it had the power to decide and standardize a technology for its market. Given its understanding and power, it did have the capabilities to meet market needs and opportunities for the country. The Korean government, as the largest shareholder of the only mobile communications operator at the time, had the power to choose technology and suppliers.

Ownership and Deployment of IP (Intellectual Property)

The Korean government did not have IPR. It recognized that Qualcomm had IPR on essential technologies for CDMA-based systems, and foreign manufacturers had essential IPR on GSM and D-AMPS. But by being the first to develop CDMA systems, it could gain IP in the methods of CDMA system implementation and applications.

The Korean government had the same situation as Qualcomm regarding the following:

- The availability of complementary products or compatibility of products/services.

- The type of technology innovation.
- The availability of alternative or substitutable technologies.

3.4.3. Korean Government's Interpretation of its Situation (IS-95)

Through the connection between Dr. Lee of PacTel and the Korean government, Qualcomm presented its CDMA technology to the Korean government in 1990 (Lee, 2001). Becoming aware of the potential of CDMA technology, the Korean government interpreted its situation in 1991 as affording great risks and opportunities.

If the Koreans themselves were able to commercialize CDMA technology successfully in the Korean market, then (1) the Korean electronics manufacturers would gain competitive advantage as a first mover in CDMA technology, and gain general competence in digital technology through developing and commercializing complex mobile systems; (2) the Korean government would protect the domestic market by providing a chance to the Korean manufacturers to supply the CDMA systems, rather than giving the equipment business to foreign companies; and (3) the Korean manufacturers might possibly be able to export their CDMA-related products if the CDMA technology were adopted in other regions.

The two examples below illustrate how committed the Korean government was to leveraging CDMA technology to create value for its electronics industry and domestic market. When PacTel (now Verizon) and Southwestern Bell (now AT&T), which were both interested in CDMA technology but had not yet developed any tangible products to commercialize, asked the Korean government whether Korea would buy CDMA systems from U.S. companies to deploy the CDMA infrastructure, the Korean government rejected the idea even though it did not know when Korea could develop its own commercial CDMA systems (Lee, 2001). Another example is the Korean government's reaction during a period when CDMA development wasn't going smoothly: Ericsson offered to deploy GSM infrastructure in Korea without requiring initial payment, and the Korean government refused Ericsson's proposal (Lee, 2001).

All in all, the Korean government's interpretation was as driven by national ambition as the German and French governments, but they faced rather different political and technological situations. The German and French governments had to work within the reality of Europe as a unified market – they needed to persuade European PTTs to buy into their technology which the other PTTs did not favor. The Korean government was in a position to control its own destiny, and Qualcomm fortuitously provide it with the technological path to move forward.

3.4.4. Korean Government's Strategies (IS-95)

Configuration of Value Network

The Korean government perceived CDMA development and commercialization as a way to boost the long term competitiveness of its electronics industry. It aimed to orchestrate the various Korean electronics companies, mobile service providers, and research institutes to coalesce into a strong value network. The research institute, ETRI (Electronics and Telecommunications Research Institute), with the backing of the government, supported Korean CDMA development and commercialization, and essentially assigned roles to the various Korean actors. Qualcomm and ETRI focused on providing the core technologies; four Korean manufacturers (including Samsung and LG) concentrated on designing and building CDMA networks, systems, and devices; and the mobile service providers were responsible for testing, evaluating, and rolling out and marketing mobile services (KIT, 2008).

The government thus sponsored, shaped, and refined the Korean 2G mobile value network for seven years until the successful commercialization of CDMA in 1996. This very substantial project took more than 1,000 manpower per year, and required huge investments (99.6 billion Korean won, which was about US$146 million at the time), funded by the Korean government and participating companies (KIT, 2008).

Formation of Standard Setting

The Korean government had the constitutional and institutional power to choose the mobile technology for the country. There was no need to establish the CDMA standard through any standard-setting organization. The task rather was to make sure that the Korean industry actors agreed to work together to make CDMA a commercially successful technology standard.

Openness of IPR (Intellectual Property Rights)

Qualcomm owned all of the essential IPR on CDMA technology. The Koreans' role was to closely work with Qualcomm to develop and commercialize CDMA. Since Qualcomm critically needed to demonstrate the viability of the CDMA technology, and since the Korean government was the first major actor willing to commit to CDMA, the Koreans had the opportunity to negotiate a good IPR deal with Qualcomm. For instance, the Korean government was guaranteed to have the "most favored" royalty terms in its license with Qualcomm. This IPR agreement gave the Korean manufacturers a cost advantage in supplying CDMA systems and devices over other manufacturers (Mock, 2005). Another example was that ETRI and Qualcomm agreed in 1992 that Qualcomm would return twenty percent of the royalties it received from the Korean manufacturers to ETRI for future research purposes (Mock, 2005).

3.5. Comparison of Actors in the IS-95 Standard

3.5.1. Comparison of Actors' Situations (IS-95)

The three actors faced similarities in three aspects of their situations: the availability of complementary products or compatibility of products/services, the type of technology innovation, and the availability of alternative or substitutable technologies. These factors are related to the relevance of CDMA with respect to the mobile industry's migration from 1G to 2G. The three actors, however, did not have the same capabilities, and their roles and concerns regarding 2G were very different – Qualcomm as a developer of technology and as a chip maker, PacTel as a mobile service provider, and the Korean govern-

ment as the leader of the Korean communications and electronics industry. As was usually the case, each of these actors had certain capabilities and power, but lacked other complementary capabilities to make them successful (see Table 4).

Qualcomm had great capabilities and essential IPR for CDMA technology, but it had limited resources and market position to commercialize CDMA. With the right technology, PacTel had capabilities to greatly increase its market. Though it had the power to choose its 2G technology, it was in no position to ensure that CDMA could become a successful technology standard. The Korean government had the power to set industry policies and to select a 2G mobile technology. It also had the power and capabilities to orchestrate the Korean industry actors. But it did not have any IPR or much technical competence in 2G technologies.

Table 4. Comparison of actors' situations (IS-95)

Actor / Element	Qualcomm	PacTel	Korean government
Organization's capabilities to meet market needs and opportunities	(D) As an innovator, Qualcomm had capabilities to meet market needs and opportunities, but did not have enough resources.	(D) It did not directly manufacture products, but as an operator, it had freedom to select a technology and could request manufacturers to provide the systems it wanted.	(D) The government did not manufacture products directly but it had power to leverage manufacturers and to standardize a technology in its market.
The availability of complementary products or compatibility of products/services	(S) All core and complementary products had to be developed.		
The type of technology innovation	(S) The CDMA technology was a disruptive innovation.		
Market position	(D) As a new entrant, Qualcomm did not have a strong position in the commercial mobile communications industry.	(D) PacTel had a strong position in its market, which was mainly the U.S. West Coast. It could select a technology and suppliers for its market.	(D) The Korean government had powerful position in its domestic market.
The availability of alternative or substitutable technologies	(S) There were other alternative technologies such as GSM, PDC and D-AMPS (TDMA).		
Ownership and deployment of IP (Intellectual Property)	(D) Qualcomm held IPR for the essential technologies related to CDMA.	(D) PacTel and the Korean government did not have IPR related to CDMA.	

[(S) Similarity, (D) Difference]

3.5.2. Comparison of Actors' Interpretations and Aspirations (IS-95)

Qualcomm believed that its CDMA technology was superior and could meet and fulfill market needs in 2G and beyond. Qualcomm aspired to make CDMA into a standard and to profit, as the owner of essential CDMA IPR, from the adoption and proliferation of CDMA. Although its situation at the time seemed difficult, Qualcomm still had many opportunities it could pursue. Qualcomm therefore interpreted its situation as calling for aggressive exploration of various avenues to push CDMA.

PacTel as an operator in the competitive mobile service industry had to be pragmatic in its choice of 2G technology. It had to balance the desire to grow its business against the cost and risk of committing to any 2G technology. It recognized the limitations of D-AMPS, but could not risk betting alone on CDMA. PacTel therefore interpreted its situation as calling for the support of Qualcomm to develop CDMA, and withheld commitment until CDMA proved promising enough in terms of generating capacity and proving the viability of its value network.

The Korean government's aspiration reflected the nation's aspiration to become economically strong, and to go beyond its traditional industries to become a major actor in the burgeoning digital electronics industry. Within the context of the Korean government's hunt for a sound development path in digital electronics, the opportunity to develop its own competence and value network in 2G mobile communications was very appealing. Assuming that the Korean government had correctly judged (via ETRI's research) that CDMA was indeed superior, and assuming that the Korean government could negotiate a good deal with Qualcomm on IPR and technology transfer, the situation the Korean government faced called for it mobilize its manufacturers and service providers to work hard together to commercialize CDMA in Korea. Thus, the Korean government interpreted its situation as (1) a great opportunity, and (2) a challenge to the whole nation to exercise its gumption.

3.5.3. Comparison of Actors' Strategic Moves (IS-95)

Below, the strategic moves of Qualcomm, PacTel, and the Korean government as based on their aspirations and interpretations of their situations are succinctly stated (see also Table 5).

Qualcomm's strategies were to (1) pressure the standard-setting organization TIA to review and approve CDMA technology as a 2G standard, (2) advocate for CDMA and persuade mobile service providers to consider, test, and ultimately adopt CDMA technology, (3) attract manufacturers by licensing its IPR to build the CDMA value network, and (4) in general, assist all interested actors in every possible way to benefit from CDMA as long as Qualcomm's core IPR strength was not compromised.

PacTel's strategies were to (1) invest in Qualcomm's development of CDMA technology, while not committing itself to adopting the technology, (2) influence TIA to endorse CDMA as a 2G standard, (3) secure the support of more manufacturers to produce CDMA systems by requesting Qualcomm to license its IPR at RAND rates, and (4) lobby other U.S. service providers to consider CDMA so as to gain acceptance for CDMA.

The Korean government's strategies were to (1) mobilize the Korean companies and research institutes to develop and commercialize CDMA, (2) structure good deals with Qualcomm concerning IPR and technology transfer, and (3) orchestrate the funding and cooperation necessary to move the ambitious project forward.

It should not be difficult to see how the strategies of three actors tightly complemented one another. Together they led to the realization of CDMA as a commercial 2G mobile technology.

Table 5. Comparison of actors' strategic moves (IS-95)

Actor \ Aspect	Qualcomm	PacTel	Korean government
Configuration of the value network	(D) Qualcomm visited the FCC to legitimate its approach and tried to persuade operators and manufacturers to support the CDMA technology.	(D) PacTel encouraged Qualcomm to develop the technology further by investing its resources. It didn't completely commit, but did evaluate the possibilities of the technology.	(D) The Korean government formed a value network with the Korean research institutes, manufacturers and service providers to develop and commercialize CDMA technology. In this way, many actors in Korea would receive benefits through this experience.
Formation of standard-setting	(D) Qualcomm had to persuade other actors by itself (de-facto). At the same time, it also persuaded the standard-setting organization TIA to approve cdmaOne.	(D) It asked and helped Qualcomm to receive approval from the standard-setting organization, TIA.	(D) Within the Korean market, it was a de-jure standard decided by the government. The Korean government formed a consortium to develop and commercialize CDMA technology.
Openness of IPR	(D) Qualcomm licensed its IPR to other manufacturers.	(D) It wanted Qualcomm to license IPR to other manufacturers.	(D) The Korean government tried to achieve the best license agreement from Qualcomm.

[(S) Similarity, (D) Difference]

3.6. Emergence of the CDMA Standard and Configuration through Actors' Interaction

Recall the emergence of the European 2G standard GSM as described in some detail in the section of GSM Standard. That Europe settled on GSM as a pan-European mobile standard was the outcome resulting from the interactions of many actors (PTTs, governments, manufacturers, and standard-setting organizations). Nevertheless, the story could be told broadly as the strategic struggle between the PTTs supporting Ericsson's narrowband TDMA proposal and the camp sponsored by German and French governments promoting an urban-centric broadband TDMA proposal, under the political and economic context of a pan-European agreement to establish the European Union.

The macro forces that underlied the emergence of CDMA was rather different. First all, there was no mandate for a unified 2G standard in the U.S. In fact, the general sentiment was to encourage open competition in the communications industry. Because of reasons discussed in the section of Background of the IS-95 Standard, the D-AMPS technology was on track to be adopted and commercialized in the U.S. Therefore, the 2G communications configuration in the U.S. would either be (A) a single standard, D-AMPS, adopted by all mobile service providers, or (B) CDMA as a competing alternative to D-AMPS. Whether case (B) would be realized depended on the strategic choice of the service providers on technology, which in turn depended on (1) endorsement of CDMA by TIA as a U.S. 2G standard; (2) the performance of CDMA technology; and (3) the ability of the CDMA value network to be competitive in 2G and beyond.

Regardless of the indecision of U.S. service providers, in 1990 the Korean government made a big bet on CDMA, following reasons given in the section of The Korean Government (IS-95). And Qualcomm, for the reasons given in the section of QUALCOMM (IS-95), was motivated to cooperate with the Koreans. It was vital to both parties to move a mobile technology from research to development and commercialization. From 1990 through 1995, there was little evidence that CDMA would turn out to be successful. But because TIA endorsed CDMA as the IS-95 standard, and because of Qualcomm's continuous promotion, the U.S. service providers did not give up on CDMA.

Qualcomm did promote CDMA to non-U.S. service providers beside the Koreans, but the early results were not encouraging. The world's first commercial CDMA network was launched in Hong Kong in 1995 using Motorola equipment. The performance of the network was deemed to be poor, which naturally cast great doubt on the true real world performance capabilities of CDMA.

Finally, after years of intensive development, the Koreans launched their CDMA networks. After overcoming the initial jitters, CDMA technology was found to be able to deliver high capacity robust service. And, as Figure 3 illustrates, the Korean mobile market (and the fortunes of its manufacturers and service providers) have grown spectacularly since then.

The success of CDMA in Korea confirmed to the U.S. service providers hedging their bets on CDMA that all their three considerations (1) to (3) stated earlier turned out to be favorable to CDMA. Thus, PacTel and other operators were motivated to adopt CDMA as their 2G technology, having seen the viability of the CDMA value network built by the Koreans and Qualcomm (and Qualcomm's partners such as Kyocera and Sony). As a small illustration of Korea's important role in the CDMA value network, and the benefit of CDMA's U.S. adoption to the Korean manufacturers, Figure 4 shows the growth of the Korean export of phones.

Figure 3. The Korean mobile communications market growth (Source: Ministry of Information and Communication in Korea [MIC])

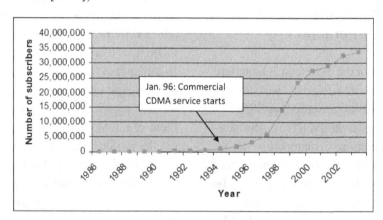

Figure 4. The Korean manufacturers' export of cellular and PCS phones (Source: Korea Association of Information and Telecommunications [KAIT])

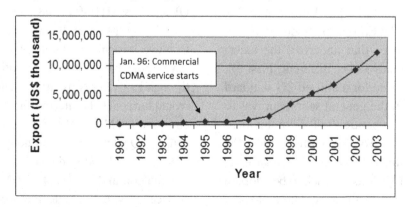

The growth and success of the CDMA networks, in the U.S. and Korea, and their potential in the rest of the world, encouraged actors who were originally more favorable to D-AMPS (e.g. Motorola and Northern Telecomm as network vendors and Ericsson and Motorola as handset vendors) to participate more earnestly in building the CDMA value network.

Thus, by the late 1990s, CDMA had emerged as a viable 2G technology with networks deployed globally and a value network composed of major global manufacturers, with Qualcomm sitting in the middle pushing further development of the CDMA technology and market while controlling the core CDMA IPR.

4. THE UNFOLDING OF THE GLOBAL 2G MOBILE COMMUNICATIONS CONFIGURATION

In this section, the unfolding of the 2G mobile communications configuration since 1996 is briefly sketched, with the goal of laying some necessary groundwork for the case study in the

next chapter on the emergence of 3G mobile communications standards.

Figure 5 gives the timeline of the development of 2G technologies up to 1996. It is clear from the Figure, and from the material in this chapter, that GSM was developed well ahead of D-AMPS and CDMA. Indeed, GSM by 1996 could be considered an entrenched and mature technology, ready to expand its dominance beyond Europe to the rest of the world. By September 1996, more than 100 mobile service providers from 58 countries outside Europe had signed the GSM MoU to adopt GSM systems (Garrard, 1998).

By late 1996 and early 1997 when operations data on the Korean CDMA network performance gradually became available, it was becoming clear that D-AMPS, though developed and adopted earlier, would not be able to compete with CDMA in performance and economy. In fact, by 1999, there were already more CDMA subscribers than D-AMPS subscribers. Although a few other markets (e.g. Mexico and Colombia) had adopted D-AMPS due to their service providers' traditionally close relationship to AT&T (Garrard, 1998), it was becoming clear that those mobile service providers outside of Europe and the U.S. who had

Figure 5. The timeline of development of the 2G standards

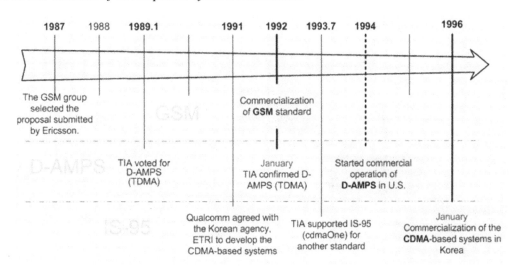

111

yet to commit would choose either GSM or CDMA as their 2G technology – the PDC technology developed by Japan's NTT, was totally isolated and held no appeal at all (see Figure 6).

Starting in 1996 and looking forward, how the future global 2G mobile communications configuration would unfold depended on how the participating actors – service providers, technology suppliers, device and equipment manufacturers, and governments – would strategically spread GSM and CDMA worldwide. It should be noted that at this point, the strategic issues were mainly that of business and technology strategies rather than standards strategy. Since this book is on standards strategy, the unfolding of the global 2G configuration was briefly discussed, with the goal of showing how the realized configuration impacted the standards strategies of the actors when they geared up to compete for dominance of 3G.

Each actor, in formulating its 2G business strategy (political and economic strategy in the case of governments), presumably examined its situation, interpreted the situation, and was guided by its aspirations in determining its strategic moves.

For example, the Chinese government by the mid-1990s recognized that it had to have a coherent 2G strategy or else it would not have communications infrastructure that could support its economic aspirations. It recognized that the choice in technology was between GSM and CDMA. When it had to make this choice in late 1995, there was little data on CDMA operations, making the performance of CDMA appear rather uncertain to China. Moreover, the Korean example of a developing country leveraging CDMA to develop its digital electronics industry was not at hand. Thus it made sense for China to make the safer bet of choosing GSM. [This is not accounting for whatever political and economic deals the Chinese and European governments agreed to behind the scene.]

We shall not survey the situation analysis of the many actors whose strategic decisions interacted to shape the global 2G mobile communications configuration. But it is noted here that the following elements (factors) and other certainties were relevant:

Figure 6. The configuration of 2G standards in mid-1990s (Note: this picture is simplified and does not show all standards adopted by every country)

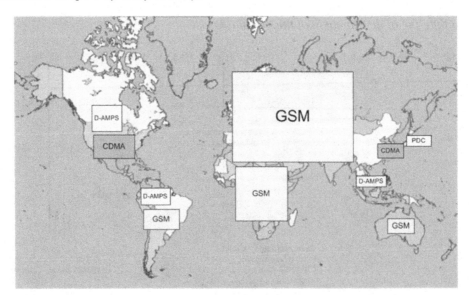

1. The support of one's own government.
2. The resources of the actor (e.g. to do vendor financing, to invest in foreign service providers, etc.).
3. The political and economic disposition of the governments in targeted foreign markets.
4. The scale and economy of GSM versus CDMA, and the relative strength and weakness of their value networks.
5. Global connectivity of GSM versus CDMA (e.g. GSM was designed for roaming while CDMA was not).
6. Transfer of technology and possible benefit to local economy.
7. Headroom of the two technologies with respect to long term deployment of mobile infrastructure.

Broadly speaking, GSM had advantages over CDMA regarding factors (1) through (5) while CDMA had advantage only in regard to factors (6) and (7). Since most countries in the late 1990s were mostly focused on the short-term benefit of quickly establishing mobile services, and since the GSM actors started their skillful international marketing of GSM years before CDMA, more countries preferred to go with GSM. Only a few countries in Latin America, because of the influence of American service providers such as Bell South and the direct investment of Qualcomm, chose to go with CDMA in the 1990s. However, in the early 2000s, with the onset of concern about the future capacity for mobile data in 3G mobile communications, Qualcomm was able to convince significant actors such as KDDI in Japan, China Unicom in China, and Reliance in India to start building CDMA networks.

The key facts about the unfolding of 2G mobile communications are three. (1) GSM dominated consistently more than 70% of the world's mobile users. (2) It was very profitable for the vendors to supply GSM equipment and for the operators to deliver GSM services. They naturally wanted to protect their turf, while the CDMA actors would scheme to crack their control of the market. (3) The strength of the GSM value network was its installed base and scale, while the strength of the CDMA value network was its technical superiority. These three facts will be seen to be a significant background condition for the development of 3G mobile standards, presented in the next chapter.

REFERENCES

Bekkers, R. (2001). *Mobile telecommunications standards: GSM, UMTS, TETRA, and ERMES*. Boston, MA: Artech House.

Calhoun, G. (1988). *Digital cellular radio*. Norwood, MA: Artech House.

Cattaneo, G. (1994). The making a pan-European network as a path-dependency process: The case of GSM versus IBM (integrated broadband communications) network. In Pogorel, G. (Ed.), *Global Telecommunications Strategies*. Amsterdam: Elsevier Science.

ETSI. (2008). IPR online database. *ETSI*. Retrieved May 8, 2012, from http://Webapp.etsi.org/ipr/

Garrard, G. A. (1998). *Cellular communications: Worldwide market development*. Boston, MA: Artech House.

Holcombe, A. N. (1911). *Public ownership of telephones on the continent of Europe*. New York: Houghton Mifflin Co..

Iversen, E. (1999). Standardisation and intellectual property rights: Conflicts between innovation and diffusion in new telecommunications systems. In Jakobs, K. (Ed.), *Information Technology Standards and Standardization: A Global Perspective* (pp. 80–101). Hershey, PA: Idea Group. doi:10.4018/978-1-878289-70-4.ch006.

Lee, W. C. Y. (2001). *Lee's essentials of wireless communications*. New York: McGraw-Hill.

Meurling, J., & Jeans, R. (1994). *The mobile phone book: The invention of the mobile phone industry*. London: Communications Week International on behalf of Ericsson Radio Systems.

Mock, D. (2005). The qualcomm equation: How a fledgling telecom company forged a new path to big profits and market dominance. New York: AMACOM (American Management Association).

Noam, E. M. (1992). *Telecommunications in Europe*. New York: Oxford University Press.

Siemens, G. (1957). *History of the house of Siemens*. Freiburg, Germany: Karl Alber.

Toutan, M. (1985). CEPT recommendations: CEPT's part in developing a homogeneous, efficient European telecommunications network. *IEEE Communications Magazine, 23*(1), 28–30. doi:10.1109/MCOM.1985.1092415.

Chapter 7
The 3G (Third Generation) of Mobile Communications Technology Standards

EXECUTIVE SUMMARY

Through the 1G and 2G mobile communications technology standardizations, the involved actors became smarter and were able to develop more sophisticated strategies. For the GSM camp, moving the whole GSM market to its favorable technology standard to control and dominate the market for the next generation mobile communications technology was extremely significant. In this situation, what could the actors in the GSM camp do? What could the CDMA camp and other actors who did not belong to either camp (actors in Japan and China) do? This chapter answers all these questions.

1. BACKGROUND OF 3G TECHNOLOGY STANDARDS

Because of the advances in digital technology and system design, all varieties of 2G technologies (GSM, CDMA, and PDC) were able to support high capacity and low cost mobile communications services. Once consumers reckoned the advantages of untethered and always-on communications at fair costs, they subscribed to mobile service *en masse*. In contrast to the niche role of 1G mobile in communications, 2G mobile unfolded to become a mass market.

DOI: 10.4018/978-1-4666-4074-0.ch007

Another difference between 1G and 2G configurations should be noted. The industrial configuration of 1G was fragmented – many local service providers providing incompatible mobile services based on a variety of incompatible technologies. In contrast, as discussed in the section of The Unfolding of the Global 2G Mobile Communications Configuration in Chapter 6, the 2G industry may be regarded as a split between just two camps, *viz.* GSM versus CDMA, with Japan's PDC being isolated and marginalized.

The third important 3G background fact to note is that 2G was focused on delivering ubiquitous voice communications. (GSM did have simple data delivery via SMS.) Now ubiquitous communications had to encompass all media – voice, data, and video – and the enabling technological advances (Internet infrastructure, IP protocol, high performance IC, etc.) were on their way to making that possible.

Thus, these three background facts together dictated that the continuing story of mobile communications was about how the GSM and CDMA camps, and the major actors outside of the two camps, competed to control and benefit from the migration of technology and market to the next generation mass delivery of multi-media mobile services, 3G.

1.1. Migration of Market from the 2G to 2.5G and 3G Arenas

By the late 1990s, as the 2G mobile communications market became more mature and saturated, the competition between companies had become more intense. This fierce competition naturally led to price erosion and loss of revenue. For example, mobile service providers set lower prices to attract customers from their competitors, which increased the costs for recruiting new customers and for maintaining existing customers. As a result, Average Revenue Per User or Unit (ARPU) deteriorated. Thus, both service providers and manufacturers were motivated to look for new sources of revenue.

However, it should be noted that the growth rate of subscribers for the 2G communications differed by region and by nation. Thus, some mobile service providers looked for new source of revenue in mature markets, while others chose to benefit from the growth of subscribers in less mature markets by investing in foreign markets. Likewise, manufacturers' profits in mature markets were being squeezed while profits in developing markets were growing strong.

Thus, the attitudes of the actors regarding the timing of migration from 2G to 3G differed according to their interpretations of their situations. For example Japan, having failed to benefit from

the worldwide 2G explosion, preferred to realize 3G as soon as possible so that it might gain shares in the global mobile market. On the other hand, China, still in the process of growing its huge 2G market, preferred to hold off 3G migration so it might have time to develop its market and its competence to compete.

With the explosive growth of the Internet in the late 1990s, the forward-looking actors in the mobile industry sensed that the future of the industry lied in mobile multimedia services delivered through the Internet. But there were huge obstacles to transforming circuit-switch 2G voice systems to packet-switch 3G multimedia systems. After discussing some of the critical technology issues involved in the section of Migration of Technology from the 2G to 3G, the section of General Background of 3G Technology Standards discusses why influencing the timing, and shaping the path of migration from 2G to 3G, were the focus of the actors' 3G standards strategies and also the focus of their subsequent business strategies.

1.2. Migration of Technology from the 2G to 3G

Given that the market was to migrate from voice services to broadband multimedia services, the driving questions were:

1. Which technologies were available to sufficiently increase signal transport to deliver broadband over a limited spectrum.
2. How easy or difficult it was, technically speaking, to migrate the existing 2G technology to the chosen 3G technology (rebuilding infrastructure if necessary); and what the migration path would be from here to there.

Given the state of technology knowledge in the late 1990s, there were only three approaches to squeezing capacity out from the limited spectrum: (1) frequency division, (2) time division, and (3) code division. Roughly speaking, since there were engineering limits to how finely the frequencies or time periods could be divided, it was not easy to extend the frequency division or time division technologies to become broadband. On the other hand, the capacity of code division technology was limited only by the designed system's ability to process pseudo-codes and to maintain power control (Mock, 2005). Since both these limits were essentially computational rather than physical, and since computational power was continuously increasing with Integrated Circuit (IC) technology (c.f. the famous Moore's Law), it was generally agreed that CDMA would be the air-interface technology choice for 3G broadband mobile.

Using CDMA as the 3G air-interface would give the CDMA camp a relatively easy and straightforward migration path because their existing 2G IS-95 (cdmaOne) systems could be "naturally" morphed to become 3G systems, all the while maintaining backward and forward compatibility. "Natural migration" meant that radical modifications in the existing systems were not required to implement the 3G technology, and the 3G air-interface could be deployed in the existing 2G spectrum.

Figure 1 shows the incremental changes that would take a 2G cdmaOne system to an interim (2.5G) CDMA 1x system, then to a full-blown 3G CDMA 1x EV DO system. (Note that radical modifications were not required.) Moreover, since the CDMA air-interface did not change, 3G chip sets and handsets could be developed through continuous innovation of design and manufacturing.

On the other hand, using CDMA as the 3G air-interface meant that the GSM camp (as well as the D-AMPS and PDC operators) would have a harder time in migration. First, because the air-interface was different, GSM operators could not implement 3G on their 2G spectrum and thus would need to acquire expensive new spectrum. Second, they would need a totally new network infrastructure (e.g. base-stations) and handsets. Third, the GSM equipment vendor would need to develop competence in CDMA technologies to pursue the disruptive innovation from TDMA to CDMA systems.

Figure 2 shows the migration path as envisioned by the GSM camp in the late 1990s. Notice that the migration would pass through the interim state(s) of General Packet Radio Service (GPRS) and/or Enhanced Data rates for GSM Evolution (EDGE) system(s), which were designed to provide data services in existing GSM frequencies, before implementing the really disruptive changes in both the air-interface and network in a new frequency band (These interim systems, however, required continuous modifications on radio and network set up.).

Figure 1. The migration path of IS-95 (Source: Update on development of IMT-2000 CDMA multi-carrier [CDMA2000] systems, Qualcomm Incorporated, presented at the 2nd meeting of the APT IMT-2000 FORUM, 6-7 December, 2001, Bangkok, Thailand)

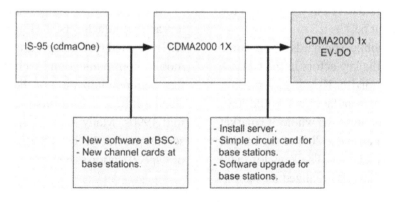

Figure 2. The migration path of GSM (Sources: Gupta, n.d.; Kalavakunta & Kripalani, 2005; Nandhini, n.d.)

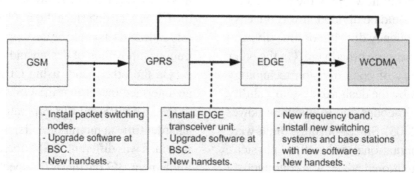

The technology migration from 2G to 3G dictated that GSM operators would ultimately have to install all new networks and introduce all new handsets. Then the (naïve?) question one might ask: why not simply junk the 2G GSM systems at an appropriate time and switch entirely to 3G CDMA? The analysis of the actors' 3G standards strategies (and business strategies) in this chapter will answer this question.

1.3. General Background of 3G Technology Standards

Although the CDMA camp held the advantage in technology and in migration path, the GSM camp had the even bigger advantage of dominating the mobile communications market. GSM had been adopted as the standard in more than 70% of the global market (Pilato, 2004). Thus, if GSM operators could be prevented from decamping to the CDMA side, the GSM camp could make its chosen 3G standard be again the dominant technology standard, regardless of how unnatural the migration path might be.

Since all 3G mobile standards were some version of CDMA, the key actors in the CDMA camp, *viz.* Qualcomm and the Korean government, believed that CDMA manufacturers would have the opportunity to supply to the whole 3G market rather than serving just under 20% of the market as in the case of 2G CDMA. The keys for them were to (1) push to have 3G realized as soon as

possible, and (2) somehow get themselves involved in the migration path of the GSM camp.

However, major actors in the GSM camp had a very different perspective. Since GSM was the dominant standard, it was continuously being diffused globally and generating great profits for the GSM camp. As long as GSM operators in the mature markets could generate ARPU through some interim technology that sufficiently met customers' desire for mobile data, and as long as the GSM camp could resist outsider pressure to roll out 3G quickly, the major mobile service providers and manufacturers in the GSM market could enjoy their positions by extending the 2G period as long as possible and moving the market to the 3G mobile communications technology arena at their own pace. The dominant GSM mobile service providers would then maximize returns on their investment in GSM systems, and the dominant GSM manufacturers would benefit from an extended product cycle and have more time to develop 3G technology.

More forward-looking than the GSM operators and manufacturers, the European Community (EC) initiated research programs for possible 3G mobile communications technologies even before commercialization of the GSM technology standard. Research results were published in the mid-1990s. Many actors, especially the mobile service providers, criticized the proposed technology specification as being too theoretical, while in truth they simply did not like the idea of rushing into 3G mobile communications (Bekkers, 2001).

While the European manufacturers were prospering with the success of GSM, the Japanese manufacturers had to satisfy themselves with their domestic market because Japan's 2G standard PDC was not adopted anywhere else. Nevertheless, the domestic Japanese market grew rapidly and soon became saturated. Therefore, to move forward, all the Japanese actors were keen on realizing 3G as soon as possible. The Japanese government did not want its electronics manufacturers to be again excluded from the future global 3G mobile communications market. It actively sponsored a study group and invited non-Japanese actors as well as Japanese manufacturers to research and coordinate plans for 3G. The Japanese government hoped that the participation of non-Japanese actors would help make the 3G technology it promoted internationally recognized.

However, the largest operator in the world at the time, NTT DoCoMo, decided to bypass the study group and in 1997, invited ten vendors, including non-Japanese manufacturers, to manufacture systems for its experimental 3G network based on WCDMA (Wide-band CDMA) technology (Bekkers, 2001). NTT DoCoMo also hoped that the participation of major manufacturers from other regions would increase the probability of its WCDMA specification being adopted as an international 3G standard.

This aggressive action of NTT DoCoMo came at a time when most actors in Europe were still resistant to EC's prompting about the development of 3G mobile technology. NTT DoCoMo's move sharply changed European actors' perspectives on 3G. The major actors in Europe began to fear that the Japanese would gain advantages as an early mover in standardizing 3G technology, so they started to actively pursue development of 3G technologies and standards that could protect and sustain the dominant market position of GSM.

The European manufacturers that provided equipment to NTT DoCoMo for its experiment (namely, Ericsson and Nokia), perhaps at NTT Do-CoMo's urging, decided to support the WCDMA proposal rather than the other proposals coming out in Europe. The European-Japanese alliance for WCDMA deliberately specified WCDMA to be incompatible with CDMA2000. The rationale for this WCDMA standards strategy seemed to be: (1) a 3G standard that was supported in the two big markets of Europe and Japan would likely become the dominant global standard; and (2) making WCDMA incompatible with CDMA2000 would block the CDMA camp out of Europe and Japan, thereby keeping the CDMA camp insignificant.

Loosely speaking, the migration of mobile communications from 2G to 3G and the unfolding of the 3G configuration up to 2008, has been shaped mainly by the struggle between WCDMA and CDMA2000. There is a third international 3G standard sanctioned by ITU, namely Time Division Synchronous CDMA (TD-SCDMA), championed by China and supported by Siemens. Its adoption, however, is restricted only to China. Therefore, the case study will be focused on the 3G standards strategy of WCDMA and CDMA2000, and TD-SCDMA in the section of The Rise of Mobile Communications in China and India will be briefly touched upon.

Lastly, it should be pointed out that the unfolding of 3G has been much more convoluted than that of 2G. When 2G mobile communications were being planned, the actors only focused on their own markets (national or regional). Their orientation was to set the 2G standard within their local domain. No one anticipated that mobile communications would explode to become a huge global market. In contrast, by 1997, every major mobile communications actor recognized that 3G business would be a global struggle. Therefore, each was much more keenly aware that its standards strategy had to be pursued in conjunction with its global business strategy, which necessarily had to adapt to changing circumstances as the 3G configuration unfolded.

Accordingly, in the case study on 3G standards strategy to follow, elements of business strategy will be brought in as appropriate. The interweaving of standards strategy and business strategy will make our modeling more complex.

2. WCDMA STANDARD

2.1. Background of the WCDMA Standard

The success of the pan-European GSM 2G standard, and the economic benefits it was bringing to Europe, demonstrated to the European Union (EU) the importance of shaping a sound communications standard. Looking beyond 2G, EU invested in research programs for planning a future pan-European communications network. The first research program (1988 – 1992), Research and Development in Advanced Communications Technologies for Europe (RACE-1), was about fixed broadcast networks in conjunction with advanced mobile networks based on a high-capacity TDMA technology, ATDMA (Advanced TDMA) (Bekkers, 2001; Holma and Toskala, 2000). The second program (1992 – 1995), RACE-2, included two projects that were concerned with developing potential technologies for a 3G mobile communications standard. One project carried the earlier ATDMA project further, and the other project, called Code Division Multiple Testbed (CoDIT), was about developing 3G communications systems based on CDMA technology (Bekkers, 2001).

Although there was an effort to concentrate resources, both projects were handed over to the next EU research program, Advanced Communications Technologies and Services (ACTS), which began in 1995. This program was later re-titled Future Radio Wideband Multiple Access System (FRAMES), and it continued research into both ATDMA and CDMA-based systems for 3G. EU allocated a large budget for FRAMES as a high priority project, and outsourced the large quantity of research to various universities and companies such as Ericsson, Nokia, and Siemens (ETSI, 1988; Holma and Toskala, 2000). To coordinate and expedite progress on the 3G projects, EU formed the Universal Mobile Telecommunications System (UMTS) Task Force to lead the research. The UMTS Task Force presented its recommendations for 3G in March 1996. The recommendations were not well received by the mobile service providers, CEPT (representing PTTs), and the standard body ETSI (Bekkers, 2001; Garrard, 1998). Mobile service providers were not inclined to invest large amounts for 3G systems implementation when their GSM markets were still growing strongly. ETSI and CEPT were offended by the fact that the UMTS Task Force was taking over their role in standardizing mobile communications technologies.

To overcome resistance, EU officially morphed the UMTS Task Force to the more participatory UMTS Forum in December 1996. In this new organizational setting, ETSI started to take the leading role in researching and shaping a pan-European 3G mobile communications technology (Bekkers, 2001). However, most other actors were still unenthusiastic about developing a 3G technology standard at the time. However, their lukewarm attitude was about to change radically due to NTT DoCoMo's aggressive movement toward 3G. The analysis of ETSI's situation and interpretation in the section of ETSI refers to this critical time period.

As mentioned, the Japanese manufacturers had not played any significant role in the global 2G mobile communications market, even though they were world-leading electronics companies. The Japanese government acknowledged this failing and was motivated to promote a 3G technology standard that would benefit Japan. It proactively set up a study group in late 1996 to study potential technologies for a 3G standard and invited both Japanese and non-Japanese manufacturers to participate. Various actors in the study group preferred different technologies for the 3G standard, based on their existing resources and capabilities gained from their experiences in 2G. For example, KDDI (merged from the companies of DDI, KDD, and IDO Corporations), a Japanese service provider that switched from PDC to IS-95 for the 2G standard, naturally preferred CDMA2000 for the 3G standard because of the easy migration from IS-95 (cdmaOne) to CDMA2000.

However, NTT DoCoMo was not interested in CDMA2000, because it would give competitive advantage to the upstart competitor KDDI. Moreover, NTT DoCoMo doubted whether CDMA2000 could become the dominant 3G standard. Therefore, NTT DoCoMo decided to experiment with its own WCDMA without going through the study group (Bekkers, 2001; Mobile Communications International, 1996). In early 1997, NTT DoCoMo ordered trial WCDMA equipment from ten manufacturers – non-Japanese companies including Ericsson, Nokia, Motorola and Lucent as well as Japanese firms including NEC, Fujitsu and others (Bekkers, 2001).

Coming from the world's largest mobile service provider at the time, NTT DoCoMo's intention to go with WCDMA could not be ignored by the non-Japanese actors. Indeed, NTT DoCoMo's sudden aggressive move prompted Ericsson and Nokia to reconsider their 3G strategies. Subsequently, both Ericsson and Nokia switched their support from ATDMA to WCDMA (Bekkers, 2001; Mobile Europe, 1997). Through its aggressive move, NTT DoCoMo became a major actor in standardizing WCDMA. Its situation, interpretation and strategy will be presented in the section of NTT DOCOMO.

The NTT DoCoMo WCDMA experiment was provocative to many then inert actors in Europe. Shocked by NTT DoCoMo's strategic move, they feared of being behind the Japanese in standardizing a 3G mobile communications technology, so they began to actively support and involve themselves in standardizing a 3G technology in Europe. The participation of many actors in the UMTS Forum made the prospect of having a pan-European 3G standard look more promising to the UMTS Task Force.

After narrowing and specifying the technological scope of 3G, ETSI encouraged actors to submit proposals that met its technical specification and set a schedule to select a proposal for the European 3G standard. To ease the migration from GSM to 3G, ETSI included backward compatibility with GSM systems as a requirement in the 3G technology specification. Therefore, the ETSI selection process was most critical for actors who were involved in the 2G communications equipment supply in Europe because it would greatly determine the migration of their future business.

The ETSI selection schedule was that: (1) proposals should be submitted by the SMG#23 meeting in June 1997 and ETSI would select and group the proposals into concept groups for further development and evaluation during the SMG#23 meeting; (2) the actors in the chosen groups would research further and present the advanced proposals at the SMG#24 meeting in December 1997; (3) one proposal would be selected through votes by ETSI's members during the SMG#24 meeting; and (4) the selected proposal would be defined, refined and optimized by June 1998 (Bekkers, 2001; Channing, 1998). ETSI's development and selection plan was very tightly scheduled, because ETSI was rushing to standardize the European 3G mobile communications technology so that European actors would not fall behind the Japanese.

At the SMG#23 meeting, proposals were grouped into five concept groups. Among them, the concept groups called *Alpha* and *Delta* were the two strongest candidates. *Alpha* was based on the WCDMA technology that evolved out of one of the projects in the FRAMES program and the research by NTT DoCoMo. *Delta* was based on a hybrid of TDMA and CDMA that evolved out of the other project in the FRAMES program, and was earlier referred to as ATDMA in this paper. *Alpha* was supported by manufacturers such as Ericsson, Nokia and Motorola, which were among the vendors of NTT DoCoMo WCDMA experiment. *Delta* was led by Siemens, Alcatel, Nortel and Italtel; they were not part of NTT

DoCoMO's experiment. This competing situation between Alpha and Delta was fatefully similar to the technology competition for the 2G standard – again pitting the Scandinavian actors against the French and German actors.

The Scandinavian actors, Ericsson and Nokia had earlier joined with Siemens to support AT-DMA, then switched to WCDMA. Why they chose to do so will be made clear by the examination of their situations. Since the situations, interpretations, and strategies of Ericsson and Nokia were very similar, they will be presented together in the section of Ericsson and Nokia.

Siemens was a leading actor for the *Delta* concept. Although ETSI ultimately did not select it to be the European 3G standard, Siemens' technology harmonized with the Chinese government's proposal was later accepted in ITU as an international 3G standard. Siemens' situation, interpretation of the situation, and strategies will be reviewed in the section of Siemens (see Figure 3).

2.2. ETSI

2.2.1. Background of ETSI

ETSI has been recognized as one of the official standard-setting organizations for Information and

Communication Technologies (ICT) in Europe since it was founded in 1988. It was also known as the organization that successfully standardized GSM technology. Based on this self- and public-recognition, ETSI was resistant to the efforts of the UMTS Task Force because it thought that the development and standardization of a pan-European 3G technology were its responsibility (Bekkers, 2001; Garrard, 1998).

The following presentation of the interpretation of ETSI covers the time when the UMTS Forum morphed from the UMTS Task Force to lead all research for the 3G standard; when NTT DoCoMo had placed an order for experimental WCDMA equipment in early 1997; and right before ETSI took over a significant role in standardizing the 3G mobile communications technology in Europe.

2.2.2. ETSI's Situation

Organization's Capabilities to Meet Market Needs and Opportunities

ETSI was not a manufacturer or an operator, but a standard-setting organization in Europe. ETSI itself did not directly have the capabilities to produce products or provide services. Therefore, the capabilities of ETSI to meet market needs and

Figure 3. The timeline of development of WCDMA standard

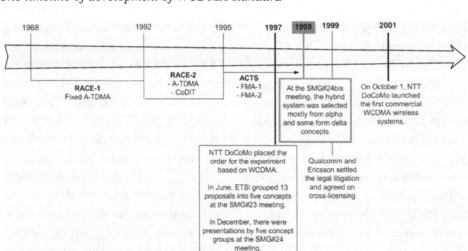

opportunities lied in its ability to standardize the right 3G communications technology for Europe, and in its ability to orchestrate its members to deliver the technology to meet market needs and opportunities.

Prior to this time, ETSI had been focusing its capabilities on diffusing GSM worldwide. In the meantime, the UMTS Task Force had taken over the role of developing a future 3G technology for Europe. But the Task Force, as a technology-focused actor, did not have the capability to steer the standardization process or to orchestrate the European operators' migration from 2G to 3G. The morphing of the UMTS Task Force to the UMTS Forum was EU's way to create an organizational situation wherein ETSI might exercise its leadership role.

Broadly speaking, the situation facing ETSI was this: (1) the manufacturers recognized the need to develop a 3G mobile broadband technology; (2) the service providers were not yet anxious to move forward; (3) the migration from 2G to 3G was going to be expensive and technically challenging.

The capability of ETSI was revealed through how well it assessed the situation (details below) and how it was able to channel the interests of the many actors towards realizing a pan-European configuration of 3G mobile technology and business.

The Availability of Complementary Products or Compatibility of Products/Services

According to the original recommendations of the UMTS Task Force, all core and complementary systems should be developed anew, because the recommended technology would not be compatible with the existing GSM systems. This incompatibility was problematic for all GSM actors – the manufacturers would have to develop new technical competence, and the service providers would have to acquire new spectrum and invest in totally new infrastructure.

The situation called for ETSI to figure out a way to develop a pan-European 3G standard while allowing the GSM actors to continue to extract benefits from GSM's continuing growth and dominance of 2G.

The Type of Technology Innovation

The technology that the UMTS Task Force recommended, from TDMA-based to CDMA-based, was disruptive innovation. The migration path the UMTS Task Force suggested implied throwing out the existing GSM systems and developing totally new 3G systems. This proposed disruptive change would require huge capital investment. In view of the fact that the anticipated demand for 3G services (data, multimedia) were then only theoretical, the European mobile service providers were naturally unconvinced of the Task Force's proposal. The situation called for ETSI to figure out migration paths from 2G to 3G that would be more acceptable to the service providers.

The Position of the Organization in the Market

As the standard-setting organization that standardized GSM and diffused it globally, ETSI had a strong reputation and influence, not only in the European market but also in the global market. Indeed, service providers and manufacturers often depended on ETSI to provide them with a comprehensive view of future markets and technologies. Moreover, as leader of the UMTS Forum, ETSI was the actor that others counted on to balance their conflicting interests and formulate a coherent plan for all. Thus, the situation called for ETSI to quickly hammer out an agreeable broad road map to 3G.

The Availability of Alternative or Substitutable Technologies

Although no 3G technology had yet been standardized, it was clear that the actors from the CDMA camp would have an easier time migrating to the

CDMA-based 3G. The CDMA actors were at the time working on specifying a 3G standard, code-named CDMA2000, for the migration from IS-95. Thus, CDMA2000 would soon become an alternative or substitutable technology for whatever technology the UMTS Forum might propose.

If the UMTS 3G technology was perceived to be very disruptive to the European service providers (e.g. requiring totally new infrastructure, etc.), they may as well switch over to CDMA2000 if it was indeed technically much better. The same would apply to GSM providers in the rest of the world. This prospect was not acceptable to ETSI because it would open the gate for the CDMA actors to come into its GSM market.

While the UMTS Task Force was making slow progress in researching 3G technologies, NTT DoCoMo was aggressively researching WCDMA and intended to developed it further (with help from manufacturers, including European manufacturers) to become an international 3G standard. WCDMA was more acceptable to ETSI, because it was partly based on the research subsidized by the UMTS Task Force, and because it would not be compatible with IS-95 or CDMA2000.

The situation called for ETSI to figure out a way to curtail the addressable market of CDMA2000, and to convince service providers that the forthcoming UMTS standard would be preferable.

Ownership and Deployment of IP (Intellectual Property)

ETSI did not own any IPR, whether essential or not. Various actors including members of ETSI already claimed many essential patents on GSM, CDMA and TDMA. It was widely acknowledged that the ownership of IP for GSM was clear and this ownership could be leveraged to protect the GSM market. On the other hand, the ownership of IP on CDMA was hotly contested, particularly between Qualcomm and Ericsson. Thus, in the setting of a 3G standard, ETSI was motivated to somehow protect the interests of the European IPR against Qualcomm.

2.2.3. ETSI's Interpretation of its Situation and its Strategic Direction

This prevailing situation pressured ETSI to realize that, as the leader of UTMS Forum, it needed to quickly standardize a 3G technology and rapidly realign value networks through the standard-setting process to catch up with NTT DoCoMo and block CDMA2000.

ETSI recognized the advantages that European actors were enjoying in the global market because of the dominance of the European 2G GSM standard. This dominance had to be protected, and should be leveraged towards dominating the 3G market. Therefore, ETSI interpreted that the 3G standard coming out from the UMTS Forum had to be seen as compatible with GSM even if disruptive technological innovation was unavoidable.

ETSI interpreted its situation as being under time pressure to proclaim a competitive, if not better, 3G standard than WCDMA and CDMA2000, and to map out a convincing migration path from GSM. Doing so would discourage GSM operators from defecting, and encourage the uncommitted to hold off joining NTT DoCoMo, or the CDMA camp.

Regarding technology, ETSI was led to believe that the European manufacturers, as dominant 2G technology vendors, also had the capabilities to develop all the necessary 3G technologies. Therefore, ETSI saw its role as hammering out an agreement among all the members of the UMTS Forum and marketing the agreed 3G standard globally.

2.2.4. ETSI's Strategies

Configuration of Value Network

ETSI's strategy in configuring value networks for 3G came from its strategic direction as discussed above – to protect and leverage GSM, and to proclaim a 3G standard as soon as possible.

First, through the UMTS Forum, ETSI rewrote the scope of the UMTS 3G technology standard to include backward compatibility with GSM systems. Although new frequency bands and a new definition of radio interface were required, the 3G switching network was scoped to morph naturally from GSM networks, thus allowed actors to preserve their investment in some of the existing equipment, such as the Home Location Register (HLR) and Mobile Switching Center (MSC) that constituted the GSM infrastructure (Bekkers, 2001).

Second, to move quickly towards fixing the standard, ETSI set a very tight schedule for the 3G standardization process:

- Selecting and grouping of the various submitted proposals into concept groups at the SMG#23 meeting in June 1997.
- Development and detailed technical evaluation of the concept groups, from June to December 1997.
- Formal presentation by the concept groups at the SMG#24 meeting in December 1997. Selection of one proposal by voting.
- Further definition, refinement, and optimization of the selected concept, to be ready by June 1998 (Bekkers, 2001, pp. 471-472).

Third, ETSI encouraged UMTS Forum members to collaborate in researching and submitting proposals. ETSI figured that through the initial collaborative research and proposal phase, and through the subsequent collaboration in the concept groups to be formed, the UMTS members would sort out their interests and align themselves to form value networks.

Fourth, ETSI set decision processes and voting rules in such a way that the CDMA camp could have little influence on the proceedings of the UMTS standard.

Fifth, the proceedings focused on getting agreement on the definition and scope of the standard rather than on the details of the technology. Although this expediting process could cause

technical problems later, avoiding getting bogged down by technical details would enable ETSI to hammer out and proclaim its future 3G standard very quickly. As long as the UMTS Forum could map out viable migration paths from GSM to 3G, ETSI could show the GSM operators and the rest of world where its dominant GSM market was heading. Thus, ETSI could convince actors to adopt its 3G standard by overwhelming them with the size of the potential market that would be inherited from the GSM market.

Formation of Standard Setting

ETSI aimed not only to set a 3G standard for Europe, but also to push it to be the dominant global 3G technology. It invited friendly non-European actors (e.g. NTT DoCoMo) to participate in the UMTS Forum. Inviting actors from outside of Europe would increase the possibility of making its 3G technology more acceptable in the world as an international standard.

ETSI also actively participated in the global standard-setting organization ITU (International Telecommunication Union) with the goal of having ITU endorse the UMTS standard as a global standard. ITU was founded in 1865 to standardize and regulate international radio and telecommunications. Although it did not play any significant role in the development of the 1G and 2G standards, since 1985 ITU had actively researched and promoted universal technology standards for global communications. ITU's goal with respect to 3G was to institute a universal agreement on the use of technology and frequency bands (1885 – 2025MHz, 2100 – 2200MHz) that would allow people to use their handsets wherever they went on the earth. The envisioned global standard for the 3G technology would be called International Mobile Telecommunication 2000 (IMT-2000).

Thus, the strategy of ETSI was to standardize the 3G technology through UMTS as the European standard. Then, it would try to standardize its European technology as IMT-2000 through ITU.

Openness of IPR (Intellectual Property Rights)

ETSI pursued a strategy of seeking open IPR. "Open" meant that a member who held IPR would be obligated to license its IPR non-exclusively to all with fair, reasonable, and non-discriminatory (RAND) terms and conditions. ETSI indicated its IPR strategy in *Annex: ETSI Intellectual Property Rights Policy* 6.1: "When an ESSENTIAL IPR relating to a particular STANDARD or TECHNICAL SPECIFICATION is brought to the attention of ETSI, the Director-General of ETSI shall immediately request the owner to give within three months an undertaking in writing that it is prepared to grant irrevocable licenses on fair, reasonable and non-discriminatory terms and conditions under such IPR to at least the following extent."

ETSI also promoted patent pooling among IPR holders on essential technologies of GSM and the potential 3G standard, so manufacturers could access to necessary patents easily by dealing with one single organization that managed the patent pool, instead of negotiating with all essential patent holders. However, patent pooling never happened due to the conflicting interests of the IPR holders.

2.3. NTT DoCoMo

2.3.1. Background of NTT DoCoMo

Although Japan's PDC technology did not gain foothold as an international standard, NTT DoCoMo successfully launched PDC systems within Japan and by the mid-1990s had become the largest mobile service provider in the world in terms of number of subscribers (Bekkers, 2001). However, the choice of PDC technology isolated the Japanese mobile communications industry from the rest of the world, and the success of NTT DoCoMo stayed within Japan.

The growing number of subscribers was burdening NTT DoCoMo's PDC systems. Also, NTT DoCoMo recognized the need to deliver mobile data service to the gadget-loving Japanese consumer base in order to increase its ARPU. So NTT DoCoMo was keen on developing a high capacity next generation mobile technology. Meanwhile, the Japanese government formed a study group for the 3G mobile communications technology standard. The following situation and interpretation of NTT DoCoMo are for the late 1996 and early 1997 time frame, right before it placed equipment orders for its WCDMA experiment.

2.3.2. NTT DoCoMo's Situation

Organization's Capabilities to Meet Market Needs and Opportunities

NTT DoCoMo had been the overwhelmingly dominant mobile service provider in Japan even though the deregulation of the Japanese communications market allowed in some competitors. Through a parallel data system of its own design, NTT launched its DoCoMo narrowband data service to complement its PDC voice service. The DoCoMo service was a great success and NTT DoCoMo's subscriber base grew tremendously. But its system capacity was becoming stressed.

With its successful experience in delivering narrowband mobile data, and with its huge profits, NTT DoCoMo was well positioned to lead the costly development and commercialization of broadband mobile technology. Although NTT DoCoMo did not directly manufacture systems, it had its own research center and enough capital to subsidize research and technology development if necessary. NTT DoCoMo could leverage not only its market knowledge but also its market power to make manufacturers produce any system that it wanted.

The Availability of Complementary Products or Compatibility of Products/Services

When NTT DoCoMo first conducted research for its next generation technology, the only known research at the time was that of the European FRAMES group. Based on the FRAMES research, NTT DoCoMo developed a wideband code-divisional 3G architecture and called it WCDMA. Since the WCDMA air interface would be completely different from PDC, and since the DoCoMo data network was rather ad hoc, NTT DoCoMo did not consider backward compatibility with its existing systems. Thus, NTT DoCoMo would need all new core and complementary systems to be developed and built for its WCDMA specification. The question for NTT DoCoMo was whether the development plans and competence of its equipment vendors would be compatible with NTT DoCoMo's WCDMA plans.

The Type of Technology and Market Innovation

As mentioned, the WCDMA technology design was a disruptive innovation. Implementing this disruptive innovation meant that NTT DoCoMo would have to junk its existing PDC infrastructure and build brand new systems. But this was feasible for NTT DoCoMo, since it had the financial resources and system capabilities to do so.

NTT DoCoMo's main concern was whether its partners and suppliers could deliver the disruptive technology quickly, because mastering necessary CDMA technologies required substantial time and resources.

The Position of an Organization in the Market

NTT DoCoMo maintained an overwhelmingly dominant market position in Japan even after Japan's privatization and liberalization of the mobile communications industry. It had the largest number of subscribers in the world. It definitely had the power to shape the 3G market in Japan.

NTT DoCoMo had tried to sell its popular DoCoMo service outside of Japan, but the effort was not successful mainly because the GSM service providers were locked into GSM and not inclined to build a separate data network. Thus, so far its influence was limited to Japan. However, if NTT DoCoMo could influence the GSM camp to adopt a 3G technology that was compatible

with its WCDMA, then NTT DoCoMo would have a better opportunity to develop its market outside Japan.

The long term threat to NTT DoCoMo's power in Japan was KDDI, a company that was formed through the merger of three service providers DDI, KDD and IDO (KDDI, 2008). KDDI's strategy was to do a disruptive infrastructure switch-over from PDC to IS-95 in 1998, and then quickly but smoothly migrate to the 3G CDMA2000. NTT DoCoMo recognized that the CDMA camp was likely to have superior technical competence. Thus, NTT DoCoMo was under time pressure to develop and implement a 3G technology quickly so as to own the market before KDDI could provide 3G services.

The Availability of Alternative or Substitutable Technologies

In late 1996 and early 1997, the UMTS Forum was far from setting its 3G standard. The specification of CDMA2000 as the 3G future of IS-95 was just beginning. The Japanese government was sponsoring research into 3G, and it intended to promote a pan-Japan 3G standard. The standard-setting process was slow because it had to account for the interests of many parties. And NTT DoCoMo's main competitor KDDI was going to propose CDMA2000 to be adopted as the standard.

NTT DoCoMo would not consider CDMA2000 as a possibility for 3G technology because its upstart Japanese competitor KDDI would in the future migrate from IS-95 to CDMA2000. NTT DoCoMo saw no reason to give up its advantages and pave the way for KDDI to succeed. Thus, if NTT DoCoMo wanted to commercialize a 3G quickly, there were no alternatives to its WCDMA.

Ownership and Deployment of IP (Intellectual Property)

The Association of Radio Industries and Businesses (ARIB), established by the Japanese government in 1995 as a standard-setting organization in Japan, published in June 1998 a survey of essential IPR ownership on WCDMA. NTT DoCoMo reported holding 43 notified patents on WCDMA technology through its own research efforts. Qualcomm claimed 163 patents, with 98 of them granted and the rest pending. AirTouch claimed 15 patents and Motorola claimed 10 (ARIB IMT-2000 Study Committee, 1998; Bekkers, 2001). The other 18 companies claimed 279 different patents. All the companies naturally claimed that their patents were essential to WCDMA. The conflicting claims of IPR ownership of essential WCDMA technology meant that patent disputes and licensing would be part of the game of commercializing WCDMA.

NTT DoCoMo was willing to license its patents to manufacturers and to trade them with other patent holders. The concern was to what extend Qualcomm had locked up the really essential CDMA technology.

2.3.3. NTT DoCoMo's Interpretation of its Situation and its Strategic Direction

First of all, NTT DoCoMo needed more capacity and better technology to meet its growing market demands and expand its popular DoCoMo data services. Second, NTT DoCoMo thought the standard-setting process in Japan was going too slowly, and would likely detract from its market dominance. Third, NTT DoCoMo believed its strong market position and rich resources would allow it to pursue a 3G mobile communications technology by itself. Doing so would give it the advantage of being the first mover, at least in Japan.

NTT DoCoMo thought WCDMA was appropriate because (1) it was incompatible with competing CDMA2000 technology; (2) it was partially based on European research so the European manufacturers could support and even help push it to be part of the European standard; (3) NTT DoCoMo held IPR on WCDMA to trade with other IPR holders.

All in all, NTT DoCoMo was keen on pushing to commercialize WCDMA in Japan as quickly as possible, and at the same time to maneuver WCDMA to be an international standard.

2.3.4. NTT DoCoMo's Strategies

Configuration of Value Network

To quickly commercialize the disruptive innovation of WCDMA, NTT DoCoMo needed to configure a whole new value network and mobilize all the manufacturers that could support WCDMA. It therefore invited all major Japanese manufacturers and also non-Japanese manufacturers with CDMA expertise – Ericsson, Nokia, Motorola and Lucent (Bekkers, 2001) – to design and manufacture WCDMA equipment for its WCDMA experiments. Qualcomm, being the technological force behind KDDI, was not invited even though Qualcomm claimed the largest number of essential patents on CDMA.

The bet was that the diverse members of this value network would have the technical wherewithal to commercialize WCDMA quickly. And, with the support of these partners, NTT DoCoMo would have a better chance for WCDMA to become an international standard.

Formation of Standard-Setting

Recognizing the slow standard-setting process in Japan and the involvement of its competitors in setting the Japanese 3G standard, NTT DoCoMo decided to go alone and establish WCDMA as a 3G standard directly through the market.

However, NTT DoCoMo aimed to actively participate in various standard-setting organizations to influence them to make WCDMA an international standard. It convinced Ericsson and Nokia to promote WCDMA technology in Europe's UMTS Forum. It participated in UMTS Forum meetings to support the WCDMA-friendly proposal submitted by its European partner vendors. As a member of ARIB (the Japanese standard-setting organization), it also submitted a 3G proposal based on WCDMA to ITU through ARIB.

Thus, NTT DoCoMo's strategy in the formation of standard-setting was two-pronged – (i) standardizing the technology by itself in the Japanese market, but (ii) going through standard-setting organizations for the global market.

Openness of IPR (Intellectual Property Rights)

As a service provider, NTT DoCoMo did not want to be locked into a few manufacturers for the WCDMA systems. It also hoped that the Japanese manufacturers would be able to supply WCDMA systems, even though non-Japanese companies held most of the essential patents on WCDMA. Therefore, its strategy was to promote open IPR. "Open" meant that manufacturers who did not have IPR on the essential technologies could still have RAND licenses from IPR holders. NTT DoCoMo had acquired a number of patents on WCDMA. Together with Ericsson's and Nokia's patents on WCDMA, NTT DoCoMo thought that its WCDMA value network could hold its ground when bargaining with Qualcomm over IPR.

2.4. Ericsson and Nokia

2.4.1. Backgrounds of Ericsson and Nokia

A brief historical background of Ericsson was given in the section of Backgrounds of Ericsson and Nokia. Here, a brief historical background of Nokia will be presented, before giving the backgrounds of Ericsson's and Nokia's 3G mobile standards strategies.

Nokia, established as a small forestry company in 1865, began its involvement in the communications industry as a manufacturer of telephone and telegraph cables in 1967 after merging with two other companies: Finnish Rubber Works and Finnish Cable Works (Steinbock, 2001). Since

then, Nokia has been actively involved in the manufacturing of commercial and military mobile radios. Nokia gained its capabilities in designing and manufacturing NMT mobile devices through Mobira Oy, which was established by Nokia and Salora Oy in 1979 (Palmberg, 2002). Nokia eventually acquired Salora Oy in 1984. Thus, through mergers and acquisitions, Nokia progressively secured its manufacturing capabilities and penetrated the mobile equipment market rapidly. As a mobile device and network supplier, Nokia grew quickly with the successes of the NMT and GSM standards (Steinbock, 2001).

As discussed in Chapter 6, Ericsson played an instrumental role in establishing GSM as the European 2G standard. As Nordic communications companies, Nokia and Ericsson had similar strategic directions and experience in the standardization and growth of the 1G NMT and 2G GSM technologies and business. And as leading global 2G mobile equipment vendors, both were invited by NTT DoCoMo to supply equipment to its WCDMA endeavor. All in all, with respect to the development of a 3G standard, Nokia and Ericsson faced similar situations and turned out to form similar interpretations of their 3G opportunities. Therefore, their 3G standards strategies will be presented together.

Due to the small size of their native Nordic markets, both Ericsson and Nokia were export-oriented companies. To supply 2G equipment globally, both Ericsson and Nokia were willing and flexible in designing and manufacturing various systems based TDMA technology, GSM or AMPS. Although Ericsson had invested significantly and gained some capabilities in CDMA technology in the 1990s, it chose not to supply CDMA-based systems (Bekkers, 2001). With their focus and success in TDMA, in the late 1980s and early 1990s as they contemplated the next generation mobile technology, Ericsson and Nokia naturally preferred developing an Advanced TDMA technology so they could utilize and extend their existing capabilities and resources (Bekkers, 2001; Mobile Europe, 1997). In the early 1990s, ambitious to extend their strong positions into the forthcoming 3G arena, both Ericsson and Nokia invested heavily in research on ATDMA technology,

However, the demonstrated technical superiority of CDMA (through IS-95 and EU-sponsored research), and the order request by NTT DoCoMo for its WCDMA endeavor, prompted Ericsson and Nokia to reconsider their strategic direction for 3G. This section will present the situations of Ericsson and Nokia at the time when NTT DoCoMo invited them to supply WCDMA-based systems for its WCDMA experiment.

2.4.2. Ericsson's and Nokia's Situations

Organization's Capabilities to Meet Market Needs and Opportunities

Ericsson and Nokia were the dominant 2G device and systems vendors. In the mid-1990s, they together owned upward of 40% of the global mobile market. In the process of growing their dominant market positions, they had built strong value networks and a full range of competences – research and development, system design, chip design, software, IPR licensing, and global manufacturing. They were generally perceived as, and they saw themselves as, the global leaders in mobile technology.

Since both Ericsson and Nokia had supplied military radios, they believed that they also had competence in CDMA since the spread-spectrum technology was originally developed for military communications. The fact that in September 1996, Ericsson sued Qualcomm for patent infringement on CDMA shows how strongly Ericsson believed that it had capabilities for CDMA.

Therefore, it was reasonable for Ericsson and Nokia to believe that they had the capabilities to shape and dominate the forthcoming 3G technology and business, and meet whatever market needs and opportunities that would arise.

The Availability of Complementary Products or Compatibility of Products/Services

The 3G ATDMA technology Nokia and Ericsson were developing was conceived to be compatible with 2G GSM technology. In contrast, the WCDMA technology specified by NTT DoCoMo was not compatible at all with any existing 2G systems, and also not compatible with the CDMA2000 specification championed by Qualcomm.

Given the rich resources and apparent determination of NTT DoCoMo, WCDMA was likely to become the first 3G technology to be commercialized, and if successful, it could evolve to become an international standard. Given that the CDMA camp had a natural migration path from 2G IS-95 to 3G CDMA2000, it was also likely that CDMA2000 would be commercialized before 3G could materialize in Europe in any form.

The risk for Ericsson and Nokia was that the global demand for 3G mobile broadband service might ramp up before Europe was ready to migrate to 3G. And if Europe's 3G standard were incompatible with WCDMA and CDMA2000, then the European dominance of the mobile market would dissipate.

The Type of Technology and Service Innovatio

NTT DoCoMo's WCDMA specification was a disruptive innovation. First, the CDMA air interface was completely different from the 2G TDMA air interface. Second, NTT DoCoMo planned on totally junking its 2G system. Third, the multimedia and location data services NTT DoCoMo aimed to deliver were fundamentally different from simple 2G voice service.

At the time, few other carriers were convinced that demand for mobile broadband service would come any time soon. The GSM operators were unsupportive of disruptive technology changes when they were enjoying significant profits from their GSM investments; they also believed that narrowband data services such as Instant Messenger was a reasonable way to add ARPU through their

GSM infrastructure. In contrast, the IS-95 CDMA operators knew that they would have a smooth migration path to 3G broadband services using the CDMA2000 standard. They therefore could time their migration according to the development of the market demand for mobile data.

This situation called for Nokia and Ericsson to figure out a way to stake their claim on the fast emerging disruptive technology and service while slowly migrating the reluctant GSM operators towards 3G.

The Position of the Organization in the Market

Nokia and Ericsson were the dominant mobile equipment vendors, and generally perceived as the global leaders in mobile technology. They therefore had great influence over the GSM operators.

Since the European GSM operators were not yet keen on 3G, they would not be the movers and shakers in setting the UMTS standard although they were voting members of the UMTS Forum. As long as a viable migration path was mapped out, they would not have a strong preference regarding ATDMA versus WCDMA. Thus, Nokia and Ericsson were in position to persuade GSM operators to support WCDMA.

The Availability of Alternative or Substitutable Technologies

Adopting the CDMA2000 standard championed by Qualcomm and the Korean vendors was not a viable alternative for Nokia or Ericsson, since it would mean giving up the benefits from leading the mobile market. In contrast, Nokia and Ericsson were invited to be leading technology partners in the WCDMA value network; WCDMA would be acceptable as the European standard if it could be morphed to be compatible with GSM. The situation called for Nokia and Ericsson to decide on the pros and cons of supporting WCDMA as the European 3G standard *versus* continuing with their own ATDMA endeavor.

Ownership and Deployment of IP (Intellectual Property)

The proper terminology for the European 3G standard is UMTS; but it is widely referred to as WCDMA because the UMTS Forum ultimately chose to adopt the wideband CDMA technology. Although confusion might arise, the term WCDMA will be used for both the WCDMA specification of NTT DoCoMo and for the UMTS standard even though they are not exactly the same.

As of March 1999, Ericsson claimed to have 46 essential patents on WCDMA technology when ETSI requested IPR holders to declare their essential patents in the Universal Mobile Telecommunications System (UMTS) proposal (Bekkers, 2001; ETSI, 2008). Interestingly, Ericsson did not declare any patents on WCDMA technology when the Japanese ARIB did its patent survey the previous year. Twenty-seven actors declared WCDMA patents to ETSI, while only eighteen did so in the ARIB survey (Bekkers, 2001; ETSI, 2008).

Nokia learned the significant value of IPR the hard way through litigation with Motorola in the 1980s concerning GSM patents (Steinbock, 2001). Before this IPR dispute, Nokia focused on developing technologies instead of filing patents, and it filed about only ten patents a year (Steinbock, 2001). After this legal battle, Nokia increased the number of patents it filed to 800 patents in 1998 and 1000 patents in 1999 to build a strong IP portfolio (Steinbock, 2001). However, when ETSI requested WCDMA patent declaration in March 1999, Nokia did not claim any patents.

Companies at the time were making conflicting claims on their ownership of essential IPR on WCDMA and CDMA. Indeed, the legal litigation about patent infringement between Ericsson and Qualcomm was still ongoing. These facts illustrated that the actors recognized more and more the significance of IPR and began to manage and leverage their IP portfolio for strategic purposes. Nokia and Ericsson therefore had to consider their patent positions versus NTT DoCoMo and versus Qualcomm when they pondered whether they should support WCDMA.

2.4.3. Ericsson's and Nokia's Interpretations of their Situations and their Strategic Direction

First, the overriding concern of Nokia and Ericsson was to protect and leverage their dominant positions in 2G in the face of possibly disruptive migration to 3G technology. In this regard, their interest was parallel to that of ETSI.

Second, it appeared imperative for Nokia and Ericsson to throw roadblocks on the advancement of CDMA2000. In this regard, Nokia and Ericsson had interests parallel to NTT DoCoMo.

Third, switching over from ATDMA to WCDMA had the attraction of handicapping other GSM equipment vendors (e.g. Siemens and Alcatel), which had been focused more on extending TDMA technology and working narrowly in the European market.

Fourth, as long as GSM operators could be constrained to migrating to a 3G technology that Nokia and Ericsson would dominate, it really did not matter much what 3G format was chosen as the standard. Nor did it appear to matter how long and costly the migration might be. Indeed, the equipment vendors could actually make more money from a tortuous 3G migration, as long as the GSM operators did not defect to CDMA2000.

Fifth, Nokia and Ericsson believed that they and NTT DoCoMo had substantial IPR in WCDMA technology. In any event, they figured that they had sufficient political and legal means to counter any excessive IPR claims from Qualcomm.

All things considered, the right strategic direction that Nokia and Ericsson perceived was to:

1. Ally with NTT DoCoMo to promote WCDMA technology as the global 3G standard.

2. Propose a modified WCDMA that was compatible with GSM to be the UMTS standard.

3. Focus on getting agreements (particularly from the operators) to adopt WCDMA in Europe without getting bogged down in the technical details of migration.

4. Help NTT DoCoMo commercialize its version of WCDMA, and in effect let NTT DoCoMo finance the cost of WCDMA development while gaining experience and piggy-back riding on NTT DoCoMo's aggressive push towards 3G.

5. Once WCDMA had been adopted as the UMTS standard, use marketing to cultivate the dogma that WCDMA would definitely be the future in both Europe and Asia, so as to block any defection to CDMA2000 and attract new operators.

2.4.4. Ericsson's and Nokia's Strategies Implementation

Configuration of Value Network

Ericsson and Nokia persuaded NTT DoCoMo to modify its WCDMA technological specification to allow backward compatibility with GSM systems (Bekkers, 2001; CommunicationsWeek International, 1997).

Nokia and Ericsson developed a version of WCDMA that was backward compatible with GSM for submission as a UMTS proposal.

Nokia and Ericsson encouraged NTT DoCoMo to participate in the UMTS Forum, to lend support to its WCDMA proposal.

Nokia and Ericsson used their reputations to convince the GSM carriers that their WCDMA proposal would enable the most sensible migration path from GSM to 3G, and lobbied for their voting support in the UMTS Forum.

Formation of Standard Setting

As systems manufacturers, it was impossible for Ericsson and Nokia to standardize a mobile communications technology in Europe without going through the UMTS Forum. Thus, the strategic plan of Ericsson and Nokia was to work hard to make their WCDMA proposal be adopted as the European 3G standard by the UMTS Forum, and then through ETSI, submit WCDMA to ITU for consideration as the IMT-2000 global standard. Concurrently, they worked with NTT DoCoMo to make WCDMA the *de facto* 3G standard in Japan, and supported NTT DoCoMo's WCDMA proposal to ITU.

Openness of IPR (Intellectual Property Rights)

Since no single actor had all the essential patents on WCDMA, all manufacturers would have to negotiate with other IPR holders for access to necessary patents. Fortunately for Nokia and Ericsson, they had vigorously filed patents in the late 1990s and built substantial patent portfolios as a basis for license negotiation with other IPR holders.

However, Qualcomm apparently held truly essential patents for CDMA technology (on rack receiver, soft handover, and power control). Since the air interface of WCDMA was partially based on CDMA technology, it seemed inevitable that Nokia and Ericsson would ultimately have to bargain with Qualcomm about licensing its essential patents for CDMA.

Ericsson believed that Qualcomm infringed on certain of its patents for CDMA, so it sued Qualcomm for patent infringements in 1996. The litigation went on until they settled with a package of cross-licensing and other deals in March 1999 (Bekkers, 2001; Mock, 2005; Westmand, 1999). But coherent IPR licensing and legal litigation strategies could not be formulated before the battle for standard-setting had been settled.

The strategies of Ericsson and Nokia around IPR were (1) building IPR portfolio and IPR management competence; (2) negotiating cross-license agreements with IPR holders; (3) forming IPR alliances; and (4) preparing to apply legal litigation and government regulations and influence to achieve better positions in negotiations.

2.5. Siemens

2.5.1. Background of Siemens

The history of Siemens is closely related to the history of German public communications. Siemens was founded in 1847 as a small company in the telegraphy business, but quickly grew by working with the German government (Noam, 1992). It would be fair to say that in the 19th century Siemens grew its business by riding the development of the German telegraphy and telephone industry. This relationship was sustained through the 20th century as well. Siemens was a major system supplier to the German PTT, Deutsche Bundespost, and by 1986, Bundespost accounted for about one-third of Siemens' revenue.

In the 1970s, although Siemens had entered new technology markets such as computing and digital electronics, it fell behind French competitors in digital development (Noam, 1992). As for mobile communications, Siemens only concentrated on the development of the German standard C450-20 and did not even participate in GSM development at the beginning. Only after GSM was adopted as the European 2G standard did Siemens participate in GSM commercialization, with help from Ericsson. Nevertheless, Siemens was not an insignificant actor in European mobile communications because of its market size, its resources, and its political relationship with the German government.

Siemens perceived the upcoming 3G mobile communications technology standardization as an opportunity to expand its role in the mobile industry. It proposed using a continuous technology innovation of GSM for the 3G standard. Below is presented Siemens' situation and its interpretation when it submitted its proposal to ETSI in 1997.

2.5.2. Siemens' Situation

Organization's Capabilities to Meet Market Needs and Opportunities

Siemens had not manufactured and supplied internationally the variety of mobile systems based on different types of standards as Ericsson and Nokia had. It had been the major communications systems vendor in Germany and its focus had been the German market.

Siemens gained capabilities and market share in 2G with the help of Ericsson. Siemens was able to grow its market along with the expansion of the GSM market. It had the capabilities to meet current market needs and the competence to improve GSM systems. However, Siemens lacked experience in the global mobile market, and its capability for CDMA technology and in managing disruptive innovation was far from secure.

The Availability of Complementary Products or Compatibility of Products/Services

Siemens had invested in researching ATDMA as a potential technology for a 3G mobile communications standard. Although ATDMA was a hybrid of TDMA and CDMA technologies, it was predominantly based on TDMA (Bekkers, 2001; Holma and Toskala, 2000). Thus, ATDMA was compatible with the GSM standard, and some GSM systems were available as complementary products.

The Type of Technology Innovation

Due to the fact that ATDMA was based on TDMA technology and therefore compatible with the GSM standard, it was basically a continuous technology innovation from GSM. The migration from GSM to ATDMA should be relatively smooth and not disruptive.

The Position of an Organization in the Market

Although Siemens joined the development of the GSM standard later than other actors, it rapidly gained capabilities in developing GSM systems with the cooperation of Ericsson and acquired the second largest market share for GSM switching systems in Europe (see Table 1). Therefore, its market position in GSM was strong, even though it did not gain a significant market share for mobile communications handsets. However, it had little influence outside Europe. It did not have the reputation of being a technology leader, and its influence on the operators was doubtful.

The Availability of Alternative or Substitutable Technologies

Because of the relative insularity of Siemens, it did not consider WCDMA or CDMA2000 as alternatives. It was focused solely on developing the ATDMA technology.

Ownership and Deployment of IP (Intellectual Property)

Siemens claimed IPR on various essential patents related to TD-CDMA that was part of the ATDMA technology when ETSI requested actors to claim essential patents during the 3G standard-setting process (ETSI, 2008).

2.5.3. Siemens' Interpretation of its Situation

Just like Ericsson and Nokia, Siemens interpreted its situation as calling for extending the dominance of GSM into the next generation mobile market. And like Nokia and Ericsson, Siemens regarded ATDMA as the most natural 3G technology for Europe. While Ericsson and Nokia changed this view and came to support WCDMA with the request for equipment from NTT DoCoMo, there was no motivation for Siemens to change its view on 3G technology. Thus, Siemens was set on developing the ATDMA technology that would continuously evolve the GSM standard (Bekkers, 2001; Vojcic et al., 1991).

Table 1. GSM market share in Europe (based on subscriber numbers in December 1996)

Supplier	Switching sub-systems		Mobile device sub-systems	
	Subscribers, in thousands	Market share	Subscribers, in thousands	Market share
Ericsson	10,297	48%	7,978	37%
Siemens	*4,426*	*20.6%*	325	2%
Nokia	3,086	14.4%	4,617	22%
Alcaltel	2,228	10.4%	2,084	10%
Lucent	515	2.4%	950	4%
Matra	443	2.1%	664	3%
Nortel	303	1.4%	0	0
Motorola	140	0.7%	2,871	13%
Others	-	-	1,938	9%

(Source: Bekkers, 2001, p. 329)

Headquartered in one of the largest countries in Europe, Siemens was disposed to forming alliances with actors from the countries that could offer large domestic markets. Siemens and these actors, with their focus on serving a few large markets, tended to prefer stable business and continuous innovations, mainly because their manufacturing infrastructures could not withstand disruptive innovation well.

In spite of the historical lesson of the European 2G battle wherein the small Nordic manufacturers won, Siemens figured that the GSM community at large, in their desire to preserve their benefits from GSM, would support its ATDMA proposal for continuous innovation of GSM to 3G.

2.5.4. Siemens' Strategies

All in all, Siemens' strategy was rather conservative and not pro-active:

Configuration of Value Network

Siemens' strategies for configuring a value network were (1) to utilize the existing value network from the GSM market and (2) form alliances with the actors that could access large domestic markets

such as France and Italy; (3) together they could increase their lobbying power to standardize ATDMA technology in ETSI (Bekkers, 2001).

Formation of Standard Setting

Siemens acknowledged ETSI's position as one of ICT standard-setting organizations in Europe. Its only option was to submit its proposal for the 3G mobile communications technology standard to ETSI, to be considered by the UMTS Forum.

Openness of IPR (Intellectual Property Rights)

According to the ETSI IPR online database (2008), "Siemens declared that it was completely committed to the IPR policy of ETSI, *Clause 6.1* for any ETSI standard relating to the ATDMA proposal and the WCDMA proposal." Thus, its simple IPR strategy was to manage its IP portfolio to license or cross-license with other actors.

2.6. Emergence of the WCDMA Standard (Round One)

This section will review how WCDMA technology was adopted by the UMTS Forum as the European 3G mobile communications standard, and subsequently endorsed by ITU as a global 3G standard.

As explained in the section of Standards Strategy in Chapter 1, the adoption of a technology standard, particularly one that will affect the prospect of businesses, requires the willing (or unwilling) agreement of the members of a community. Agreement is possible only when the members, after sorting out the pros and cons of the value networks entailed by the competing proposals, collectively endorse (or are forced to accept) a particular proposed standard. The UMTS Forum, and the European (and global) communications market at large, was a huge community. It is impossible to trace in detail how its numerous members maneuvered and reacted to each other's strategic and tactical moves in their efforts to promote and protect their interests. However, the path of emergence of the UMTS standard was largely shaped by the five major actors. They came to the game with coherent strategies that would enable and channel others to judge wherein lay their interests, and thereby sort themselves out – for or against – with respect to the proposed future value networks.

As mentioned, these five actors deliberately developed technologies that were not compatible with IS-95 and CDMA2000 so their competitors from the CDMA camp would be handicapped in the global migration from 2G to 3G. How the CDMA camp responded to these moves will be discussed in the next section. This section is focused on the UMTS Forum members and their strategic conflicts and reconciliations that ultimately led rise to the European/Japanese WCDMA standard.

The standards strategies of the five major actors – ETSI, Nokia, Ericsson, NTT DoCoMo, and Siemens – have been explained in detail in the previous sections. By considering how these strategies complemented or obstructed one another in soliciting the support of the UMTS members, one can see clearly the rationale of the emergence of the WCDMA standard in the UMTS Forum and subsequently in ITU.

Various research consortia submitted in total thirteen 3G UMTS proposals to ETSI, all backward compatible with GSM as required, for consideration prior to the SMG#23 meeting in 1997. The research consortia were formed by alliances of various manufacturers and mobile service providers. Their technical proposals, if done well, would strategically combine technical efficacy with a vision of the future value network that others could rally around.

At the SMG#23 meeting, ETSI sorted the thirteen proposals into five "concept groups" for further study and consideration. Table 2 presents the five concept groups.

As the responsible standard-setting organization, ETSI strived to shepherd the formation of the "right" future value network for 3G mobile through the standard-setting process. It encouraged actors to participate in the standardization process, and tried to minimize barriers for certain selected actors to join its value network. For example, ETSI provided Asian mobile service providers active in ETSI as observers the opportunity to become associate members with full voting rights (Bekkers, 2001). Another example was that ETSI asked Ericsson and Qualcomm to settle their ongoing litigation over patent infringements.

According to a poll at the SMG#24 meeting in December 1997, the members of ETSI favored *Alpha* and *Delta*. However, neither of them gained enough votes to be selected even after the second round, when members voted to choose between the two strongest candidates (Bekkers, 2001; ETSI, 1999). A candidate proposal needed at least 71% of the votes to be selected. *Alpha* gained 58.45% of the individual weighted votes, while *Delta* had 41.55% in the second voting round (Bekkers, 2001; ETSI, 1999). Since no proposal received the necessary percentage, the selection of a proposal for the 3G technology standard was postponed to the following SMG#24bis meeting. This meeting was to be held a month later, because ETSI wanted to submit a unified European proposal to ITU by June 1998 (Bekkers, 2001).

Table 2. Five concept groups in ETSI

Concept	Also known as	Supported by
Alpha	WCDMA, CoDIT, Frames FMA-2	NTT DoCoMo, Ericsson, Nokia, Lucent, Motorola, Fujitsu, NEC, Panasonic, Telecom Italia Mobile and ARIB as an observer
Beta		Sony, Telia and Lucent
Gamma	ATDMA, Frames FMA-1 without spreading	Philips, Nokia and France Telecom
Delta	TD/CDMA, Frames FMA-1 with spreading	Siemens, Alcatel, Nortel and Italtel. Motorola, Bosch and Sony joined later.
Epsilon		Vodafone and Salbu R&D

(Source: Bekkers, 2001, p. 473; Channing, 1998)

Although no final selection was made, the voting result in SMG#24 was very important. It showed that there were only two strong candidates, and one of them was therefore expected to eventually inherit the GSM market. And all the UMTS actors had to sort out their alliances and stake their positions with respect to the two proposals within a month.

Thus, it was useful to consider how actors might form value networks around these two 3G "concepts". The concept *Alpha* was based on WCDMA and was supported by the group of international actors that had participated in NTT DoCoMo's experiment—Ericsson, Nokia, Lucent, Motorola, NEC, etc. The concept *Delta*, based on ATDMA, was led by Siemens, Alcatel and Italtel, all companies from European countries with large populations.

It should be recalled that Ericsson and Nokia persuaded NTT DoCoMo to change its WCDMA technological specification to become compatible with GSM, thus making WCDMA attractive to the European actors. The European operators did sense that a CDMA-based technology would be technically superior; their worries were about disruptive migration. The international flavor of WCDMA also appealed to those operators with international ambitions, e.g. Vodafone. On the other hand, Siemens' conservative strategy suggested a continuous innovation to the GSM standard, shaping a value network with the large European countries (Germany, France, and Italy) at the center.

The competition between these two concepts looked similar to the competition between the two proposed value network configurations during 2G GSM standardization a decade ago (Bekkers, 2001). One was forward-looking, inclusive, and international; while the other was conservative, exclusive, and regional. The fact that in both cases one side was led by the Nordic actors and the other by the alliance of larger countries was not totally coincidental. It strongly reflects how deeply one's situation and orientation and aspirations shape and constrain one's strategy.

If a unified UMTS 3G standard were to emerge, the ETSI members had to show their preference regarding the two competing value network proposals, and ETSI as an organization had to adjudicate and harmonize the conflicting interests. The following section will continue describing how that happened.

2.7. Emergence of the WCDMA Standard (Round Two)

During the month between the SMG#24 and SMG#24bis meetings, the participating actors such as Ericsson, Nokia, and Siemens aggressively promoted their competing proposals and criticized their competitors, while many actors urged them to reconcile (Bekkers, 2001; GSM MoU Association, 1999).

In the second round of voting during the SMG#24bis meeting (Paris, January 28-29), *Alpha* received the most votes – *Alpha* (61.1%) *Delta* (38.7%), *Gamma* (0.2%), and *Beta* (0%). But *Alpha* still did not have enough votes to be officially selected as the standard (Bekkers, 2001; ETSI 1998). The politically expedient discussion immediately commenced – to come up with a hybrid system that would accommodate both Alpha and Delta concepts. NTT DoCoMo, attending the meeting as an observer, confirmed that it would support the hybrid system (Bekkers, 2001). ETSI, following its strategy of "push for the standard and forget implementation," naturally encouraged acceptance of the hybrid proposal as the UMTS standard. However, it was easy to see that the concept from the Alpha proposal dominated the specification of the hybrid proposal (Bekkers, 2001). This hybrid proposal was a politically expedient reconciliation between two sides rather than a true resolution of the technical differences.

The outcome of the UMTS standard-setting process implied that the value network configuration and migration path as envisioned by Ericsson, Nokia and NTT DoCoMo would be the strategic future towards which the GSM camp and NTT DoCoMo would strive.

After settling on a blueprint for this strategic future, the actors began to discuss forming a consortium that would facilitate further research and definition of technology specifications during the remainder of the SMG#24bis meeting. ETSI eventually formed the 3rd Generation Partnership Project (3GPP) as an umbrella organization for all actors that supported WCDMA. The 3GPP organization was essentially formed by the actors from the original WCDMA alliance to reinforce their envisioned value network, and to coordinate their activities in promoting and commercializing WCDMA as a global 3G standard. Starting out as a European organization, it soon became a

global organization – TTA from Korea quickly joined 3GPP, and ARIB from Japan joined in December 1998.

This was how the European 3G standard and the European vision of the future mobile value network emerged. With European 3G standardization completed, ETSI submitted the WCDMA standard under the name of "UMTS Terrestrial Radio Access (UTRA)" to ITU for consideration as the 3G global standard IMT-2000. This UMTS submission was essentially identical to the WCDMA proposal submitted by ARIB (the Japanese standard-setting organization). This ARIB proposal in turn was basically the technology NTT DoCoMo specified for its 3G endeavor. The WCDMA standard specified by ITU in 1999 as one of the global 3G technology standards represented both the European UTRA version of WCDMA and the Japanese version of DoCoMO's WCDMA.

The European mobile service providers did not play a significant role in standardizing a 3G technology. They were too busy with serving the increased number of subscribers in existing GSM systems. From their perspective, all they wanted was a global 3G standard towards which they could migrate their GSM systems cost-effectively. They essentially relied on the European official standard-setting organization ETSI, that had been so successful in setting the 2G standard, to hammer out a 3G standard for them. This was why the major vendors of GSM systems such as Ericsson, Nokia, and Siemens dominated the 3G standard-setting process in ETSI. Implicitly, the service providers trusted that their vendors would have their interests at heart and would devise the right 3G technology and migration path for them. Thus, they did not influence the 3G standard-setting process as much as they could, and quite blindly followed the strategies of ETSI and their major vendors. Indeed, WCDMA was known as a "vendor-driven" standard. As it later turned out, the interests of WCDMA vendors and service providers were not exactly parallel – for example, the vendors' competence in CDMA technology was not as good as they projected, and the cost of WCDMA implementation and migration turned out to be much higher. The conflicts between the WCDMA vendors and service providers would be a very significant factor in the subsequent unfolding of the global 3G industrial configuration.

Next, the 3G standards strategy of the CDMA camp will be considered.

3. CDMA2000 STANDARD

3.1. Background of the CDMA2000 Standard

As described and analyzed in Chapter 6, Qualcomm, the Korean government, and other actors were able to establish IS-95 (cdmaOne) as a 2G mobile communications standard in the period from 1988 through 1996, and through commercialization of IS-95 showed the world that CDMA was an excellent technology for mobile communications.

Once CDMA technology had been proven, the actors in the CDMA camp, particularly Qualcomm and the Korean CDMA vendors, wanted to commercialize and diffuse CDMA as widely as possible. Unfortunately for the CDMA camp, CDMA came on the scene too late. By 1996, the GSM standard had already proliferated and become entrenched around the world, and it appeared that IS-95's market share would be unlikely to ever exceed 20%. Given this situation, the actors from the CDMA camp realized that CDMA technology would continue to be marginal as long as the global mobile communications market remained in the 2G stage of digital voice service (Mock, 2005). For them, the way out of this predicament was to move the mobile technology and market to the 3G stage of broadband multimedia (voice and data) communications as soon as possible, so that they would have the space to compete and win.

For the anticipated migration from 2G to 3G, the CDMA camp had several advantages: (1) for

mobile broadband, CDMA was a better technology than extensions of TDMA, with regard to both the efficiency and flexibility of the use of the scarce resource of spectrum; (2) the migration from 2G IS-95 (i.e. cdmaOne) to the future 3G CDMA was fairly smooth, and therefore could be implemented speedily to drive the migration of the global market; (3) Qualcomm owned the essential IPR on CDMA technology. On the other hand, the CDMA camp had the huge disadvantage that the GSM camp controlled close to 80% of the mobile market. The GSM camp could choose to use a less efficient technology for its 3G standard – attempts to handicap customers in order to retain profits were hardly unusual! Even if the GSM camp chose to use CDMA technology for 3G, it could specify its 3G standard to be incompatible with that of the CDMA camp, and engage other tactics to block the CDMA actors from encroaching on its turf. Moreover, since the migration from 2G GSM to 3G CDMA would be disruptive and costly, even if it came to be, it would be slow.

Given these pros and cons, the challenge for the actors in the CDMA camp was to figure out their strategies for establishing 3G mobile standards that would facilitate the pursuit of their business strategies for profit and power, necessarily as opportunists and attackers.

Qualcomm and the Korean government were the leaders in shaping the standards strategies of the CDMA camp. In the sections below, their interpretation of their situations and their setting of their standards strategies, in the 1996 to 1998 time frame, will be described and analyzed (see Figure 4).

3.2. Qualcomm (CDMA2000)

3.2.1. Background of Qualcomm (Coming Out of its 2G Success)

How Qualcomm led the development and commercialization of CDMA as a 2G mobile technology was described in the section of QUALCOMM (IS-

Figure 4. The timeline of development of CDMA2000 standard

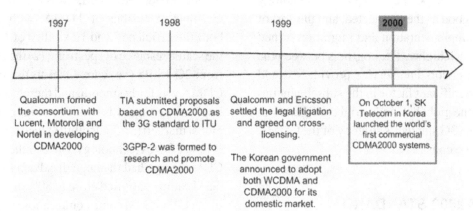

95) in Chapter 6. As the lone leader and promoter in the beginning, in order to make 2G CDMA a reality, Qualcomm had to do everything – research and development, system and network development, chip set design (fabrication of chips was outsourced), even handset design and development. In the early 1990s, the Korean manufacturers, under the stewardship of the Korean government, began to develop their competence in these areas. And as IS-95 systems began to look more commercially viable, vendors such as Lucent and Motorola were attracted to become CDMA network vendors, and manufacturers such as Kyocera were attracted to begin supplying CDMA handsets.

Qualcomm's disposition as a company is to be a vendor of technology rather than a manufacturer – it did start out as a technology consulting firm and a government (mostly defense) contractor (Mock, 2005). Its fundamental business model was to: (1) develop and patent technologies; (2) generate revenue by licensing; (3) invest strategically to develop demand for its technology, and (4) control the high-margin segment of the supply chain (for mobile, it is the design and supply of the core IC) and outsource everything else (Since 1998, Qualcomm's business has been organized by segments, with the three core segments being Qualcomm Technology Licensing, Qualcomm CDMA Technologies [which does the IC], and Qualcomm Strategic Initiatives [Mock, 2005].).

Qualcomm believed that CDMA technology was superior and the only sensible technology choice for 3G mobile. Since it believed that it owned the essential IPR on CDMA technology, its business model would generate tremendous profit if (1) 3G would indeed be based on CDMA technology, and (2) somehow Qualcomm could have access to the whole mobile market.

This background understanding of Qualcomm will be helpful in assessing Qualcomm's standards strategy.

3.2.2. Qualcomm's Situation (CDMA2000)

Organization's Capabilities to Meet Market Needs and Opportunities

Qualcomm, as the master of CDMA technology and the leader of the CDMA value network, had the capability to lead the end-to-end development of 3G technologies and systems. It had the capability to supply the ICs for building 3G networks and handsets to meet the future global market demand for broadband mobile services. The challenge for Qualcomm was market creation and access.

The Availability of Complementary Products or Compatibility of Products/Services

The future 3G standard, if based on CDMA technology, could readily be specified to be backward compatibility with existing IS-95 systems. With such a 3G specification, the products and competences of existing CDMA vendors would be compatible with 3G development. However, 3G CDMA would not be compatible with existing GSM systems. Even if the GSM camp chose to base their 3G specification on CDMA technology, in all likelihood the specification would not be compatible with the 3G specification set by the CDMA camp.

The Type of Technology Innovation

The future 3G CDMA technology was going to be a continuous innovation from 2G CDMA for IS-95 developers and adopters. But 3G CDMA would be a disruptive innovation for GSM developers and adopters. This meant that if 3G was going to be based on CDMA, then CDMA vendors had a technical advantage. Indeed, given the technical difficulties of making CDMA work, GSM vendors might not have sufficient competence in CDMA (even though they claimed to have such) to make their version of 3G CDMA work well.

The Position of an Organization in the Market

Since 2G CDMA had less than 20% of the global market, Qualcomm's position with respect to the global market was not strong. However, Qualcomm's power within the CDMA camp was very strong, if not dominating.

Qualcomm had had virtually zero influence on how NTT DoCoMo or the European actors would choose their 3G technology. However, as things turned out, by specifying WCDMA, they all committed to base their 3G on CDMA technology. To the extent that Qualcomm really controlled CDMA technology, and to the extent that Qualcomm would be able to leverage its expertise, Qualcomm could be in a very strong position to dominate the 3G technology market.

However, Qualcomm as yet had little influence over the GSM actors within their GSM turf. And Qualcomm had a bad reputation among GSM operators as a brash upstart that had barged into the gentlemen's club of communications (Mock, 2005). Therefore, Qualcomm's market power in 3G would depend greatly on how it strategically built (or rebuilt) its relations with others.

The Availability of Alternative or Substitutable Technologies

Prior to 1997, Qualcomm had done much R & D on 3G. In 1997, Qualcomm joined with Lucent, Motorola and others to form the CDMA2000 consortium to fine-tune the Qualcomm design into a public 3G specification. By 1997, NTT DoCoMo had started its aggressive push to establish WCDMA as the world's first commercialized 3G technology. NTT DoCoMo invited Nokia and Ericsson to partner in its effort. That alliance ultimately (as described in the section of WCDMA Standard) led to a version of WCDMA specified as the European 3G standard towards which GSM operators would be shepherded in their future migration from 2G.

Thus, although in theory CDMA2000 could be a substitute for WCDMA, in fact the future penetration of CDMA into GSM turf looked highly unlikely. Worse yet for the CDMA camp, given the choice between CDMA2000 and WCDMA, the existing PDC operators in Japan (NTT DoCoMo was the leading one), and TDMA operators in the U.S. (such as AT&T), and the uncommitted operators in the rest of the world might go with the herd and side with WCDMA.

Ownership and Deployment of IP (Intellectual Property)

Qualcomm believed, and it turned out indeed to be true, that it owned most of the essential patents on CDMA technologies (Mock, 2005). Since competitors would rather fight than pay for IPR, nasty and extended IPR challenges to Qualcomm's IPR seemed unavoidable. On the other hand, no company could *legally* proceed with commercializing 3G unless it obtained a CDMA license from

Qualcomm. This gave Qualcomm a significant advantage in possibly bargaining its way into the 3G WCDMA market.

3.2.3. Qualcomm's Interpretation of its Situation and its Strategic Direction (CDMA2000)

The situation that developed in 1996-1998 was a mix of good and bad news for Qualcomm. The good news was that all emerging versions of 3G mobile standard would be based on CDMA technology, from which Qualcomm could profit. The bad news was that its competitors had set up and would continue to set up all kinds of roadblocks to prevent Qualcomm and its partners from going beyond their existing home turf.

Unlike the case of 2G, where actors competed to establish the form and domain of the 2G mobile standard, it was clear very early on that there would be two or three 3G standards – one from the GSM camp, one from the CDMA camp, and perhaps a third one from Japan. As it turned out, with the alliance of NTT DoCoMo with the Europeans, there was only going to be two competing 3G standards, CDMA2000 *vs.* WCDMA. (The aspirations of the Koreans and the Chinese in setting their own 3G standards simply did not seem technologically or economically viable.) Thus, standards strategies took on the emphasis of exploiting the standards, in the context of technology diffusion, to further one's business strategy.

In hindsight, Qualcomm's strategy appeared to be an artful exercise in maneuver warfare – maneuver on multiple fronts, and the dynamic interaction of frontal and oblique moves (e.g. John Boyd's Observe-Orient-Decide-Act, 2008; Sun-Tzu's The Art of War translated by Sawyer, 1994).

On the front side: (A) *Defend Turf*: Help partners to further develop the 2G CDMA market in the U.S. and Korea, and prompt them to move to CDMA2000 as soon as possible. (B) *Extend Base of Operation*: Make strategic investments to develop new 2G markets. In particular, maneuver

to help set up IS-95 operators in the two huge markets of China and India to compete with the existing GSM operators there. (C) *Attack Enemy at Home*: (1) Drive KDDI to migrate to 3G very quickly to attack and disrupt NTT DoCoMo's WCDMA plan. (2) Develop GSM 1x, which was a 2.5G technology with CDMA air interface and GSM-MAP network, to attract GSM operators to defect or at least cause them to question the official WCDMA migration path. The GSM 1x technology could sit on existing GSM systems to communicate with CDMA2000 handsets. (3) Develop CDMA450 targeted at the Nordic and Eastern European countries so as to penetrate Europe at its border. CDMA450 is a family of IS-95 and CDMA2000 systems that is deployed in 450 MHz.

On the technology side: Speedily develop superior IC for both CDMA and WCDMA, so its vendor partners could push the CDMA2000 market and get set to push and penetrate the WCDMA market; and develop full suites of patents on CDMA technologies in addition to its core essential patents.

On the oblique side: (A) by pushing CDMA2000 against WCDMA on multiple fronts, expose the weakness of the specification of WCDMA. (Recall that WCDMA was deliberately designed to be different from CDMA2000, therefore its technical specification was prone to be sub-optimal in terms of engineering, for example, choosing the more complex asymmetric signal timing at base stations just to be different from Qualcomm's use of symmetric signaling. Moreover, to expedite UMTS agreement, WCDMA was specified for Europe without sufficient engineering testing. Thus, WCDMA commercialization required lots of revision and fixing that directly related to cost and speed.) This would naturally sow discord among the WCDMA actors. In particular, the operators would begin to blame the vendors who pushed the 3G standard on them, and cast doubt on their 3G competence. (B) Qualcomm might then come in to play the role of the good guy, helping operators fix up the WCDMA standard and build their WCDMA networks. Then, Qualcomm could bring in the wider selection of 3G handsets running on Qualcomm's advance WCDMA ICs through their Korean and Japanese partners.

Qualcomm hoped that the above combination of strategic maneuvers would enable it to first address, then dominate the whole emerging 3G market. In any event, as long as 3G went forward, Qualcomm was sure to collect very significant revenue from IPR licensing (in the billions of dollars).

3.2.4. Qualcomm's Strategies (CDMA2000)

Configuration of Value Network

Qualcomm's strategies for building value networks was straightforward: (1) strengthen the existing value networks. For instance, it cultivated more and better partners, like setting up multiple partnerships with foundries for IC manufacturing. (2) Extend the value networks to support and accommodate the multi-faceted, multi-region strategic maneuvers. For example, it partnered with Japanese manufacturers to supply NTT DoCoMo with WCDMA handsets. It also developed CDMA Mobile Local Loop (WLL) as a cheap 2G technology for Indian partner operators so as to enter the huge Indian market.

Formation of Standard Setting

Qualcomm encouraged IS-95 adopters to participate in various standard-setting organizations to promote CDMA technology in general and CDMA2000 in particular. Qualcomm and partners submitted the CDMA2000 specification to TIA (the American standard-setting body in telecommunications) to be officially endorsed as a 3G mobile standard.

Through TIA, CDMA2000 was submitted to ITU (International Telecommunication Union) for consideration as a global 3G standard, in competition with WCDMA.

Qualcomm tried to join ETSI in order to introduce some compatibility between CDMA2000 and WCDMA. But, ETSI did not allow Qualcomm to become a member because it was from outside of Europe. So, Qualcomm founded a European subsidy just so it could become a member of ETSI. However, Qualcomm's voting power was very low because a company's vote in ETSI was weighted by its market share in Europe, and Qualcomm had virtually zero market share in Europe at the time.

Openness of IPR (Intellectual Property Rights)

To participate in the Japanese WCDMA market, Qualcomm had to have access to certain necessary WCDMA IPR. For the European WCDMA market, to develop a dual- or triple-mode chipset that would be compatibility with both 2G and 2.5G GSM, Qualcomm had to have access to all essential IPRs on the GSM standard. Different companies held different essential IPR on the technologies for the GSM standard (Bekkers, 2001; ETSI, 1996). Thus, Qualcomm had to negotiate license or cross-license agreements with these IPR holders.

Business IPR licensing is a key revenue-generating component of Qualcomm's business model, which was not the case for large manufacturers such as Nokia. Therefore, it would be unwise for Qualcomm to play the usual cross-licensing game with the manufacturers. Qualcomm chose a strategy of being a tough collector:

When Europe was heading toward specifying a version of WCDMA that would not be compatible with CDMA2000, Qualcomm declared that it would not license its CDMA patents to manufacturers planning to produce WCDMA systems unless certain conditions which Qualcomm deemed reasonable were met (Bekkers, 2001). These conditions included: "(1) [ETSI and its members] work toward a single, converged 3G standard; (2) the standard be made compatible with both European GSM-MAP and American ANSI-41-based network; and (3) the technical specifications be chosen clearly on the basis of proven superior performance, features, or cost" (Mock, 2005, p. 205). Later on, Qualcomm retreated from this strong demand.

Qualcomm had to sound tough at that point in time since Ericsson had initiated in 1996 litigation of patent infringement on CDMA, and the litigation was not resolved until late 1998. Ultimately, with the prompting of ITU and the court's adverse hearing against Ericsson, Qualcomm, and Ericsson settled their suit, and Qualcomm retreated from its "extreme" licensing position.

Qualcomm always intended to collect fully on its essential CDMA patents and would not cross-license them away. The difference of essential *vs.* non-essential technology may be understood as the difference between a recipe and the particular method to make the sauce that the recipe calls for. Qualcomm was ready to cross-license its CDMA technology "sauce-making" IPR for GSM and WCDMA IPR, but insisted on others paying in full for the CDMA "recipe". In fact, Qualcomm was strategically prepared to spin out its IC-making division, QCT (Qualcomm CDMA Technologies), and assign it its non-essential IPR so it could horse-trade and cross-license with the manufacturers; in the meantime, the slimmed down Qualcomm could keep and license the core CDMA IPR, suing infringers if necessary (Mock, 2005).

3.3. The Korean Government (CDMA2000)

3.3.1. Background of the Korean Government (CDMA2000)

In the late 1980s, the Korean government chose CDMA technology as the vehicle for the development of the Korean mobile communications market, and also as a vehicle for driving the development of Korea's electronics industry. The efforts of the Korean actors (government, research institutes, operators, and manufacturers) bore fruit in 1995-1996 with the successful commercializa-

tion of the 2G IS-95 mobile services, with all the network equipment and handsets designed and manufactured by Korean companies. Since then, Korea has grown to become the world's most advanced mobile market, and Korea's electronics industry has become highly sophisticated and successful. Thus, the Korean government achieved what it had wished in the late 1980s.

The Korean mobile communications manufacturers (Samsung, LG, and other second tier manufacturers) gained their competences and capabilities through the development and commercialization of the Korean CDMA systems. They grew first by supplying the local market, then by exporting to the growing CDMA market in the U.S. Since then, they have emerged as world-class mobile handset manufacturers.

Very early on, the Koreans recognized that the CDMA market was going to be much smaller than the GSM market. Thus, they smartly leveraged their capabilities gained from developing CDMA systems to export handsets for GSM markets. Indeed, since 1998, Korea's volume of GSM handset exports has been greater than that of CDMA handsets (See Figure 5).

Based on its experience from the late 1980s to the mid-1990s, the Korean government learned three things: (1) how pushing technology development could benefit the economy as a whole; (2) how technology standards could determine the size of markets and profit potential; (3) how ownership of IPR could add leverage to a company's market power. The Korean government's strategy for mobile communications henceforth was motivated by these lessons.

Aspiring to gain the benefits that owning a technology standard could bring to the nation, the Korean government encouraged the Korean actors to develop next generation mobile technologies that could form the basis of a global 3G standard that incorporated Korean IPR. In 1988, the Korean government established the Korean standard-setting organization TTA (Telecommunications Technology Association). Through TTA, the

Figure 5. Korean exports of mobile handsets (1998-2002) (Source: International Cooperation Agency for Korea IT, Monthly IT Export – 2003/10)

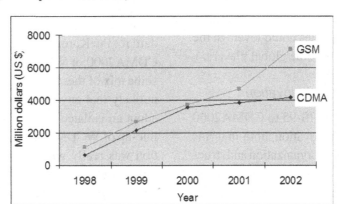

Korean government submitted two proposals to ITU for consideration as IMT-2000, the global 3G mobile communications technology standard to be specified by ITU. There was, however, little hope that the Korean proposals would prevail at ITU over their competitors, WCDMA and CDMA2000, which were championed by actors (such as NTT DoCoMo, ETSI, Qualcomm, Nokia, Ericsson, etc.) that had control over much larger markets and much stronger portfolios of IPR.

With its own 3G proposal likely to be rejected, and the world's mobile market likely to be split between WCDMA and CDMA2000, the Korean government had to assess its situation and map its strategies. Its interpretation of the situation, between the time it submitted its proposals to ITU in 1998, and the time it set 3G standards for the Korean markets in 1999, is discussed below.

3.3.2. Korean Government's Situation (CDMA2000)

Organization's Capabilities to Meet Market Needs and Opportunities

The Korean government did not directly manufacture products. It saw its responsibility as determining strategies and setting policies to mobilize and orchestrate the various Korean actors to meet market needs and potentials. It had to map out strategies for both the growing 2G market and the future 3G mobile market as well as for both domestic and international markets.

By 1998, Korea clearly had the capability to be a global supplier of 2G handsets, for both GSM and IS-95. For 3G handsets, Korea definitely had the capability to build them. But since Korea was only beginning its capabilities to design and fabricate complex ICs, it was unlikely that Korea could supply its own 3G chips to put into the handsets any time soon.

Korea's network vendors had the capability to build 2G mobile networks even though they had not yet ventured out of Korea. By building 3G networks in Korea, they would be capable of building next generation networks worldwide.

With its success in the 2G CDMA standard, the Korean government was justified in believing that it was capable of leading Korean companies to succeed in 3G and beyond. It had the political power and economic influence to orchestrate the Korean companies to pursue a coherent 3G strategy.

The Availability of Complementary Products or Compatibility of Products/Services

With Korea's industry so export-oriented, the question was not how other products and services would be compatible with Korea's in the 3G stage. Rather, the question was how the products and services of Korea yet to be developed would be compatible with the market needs of others, so that the Korean companies could leverage the future growth of 3G to become global players.

The Type of Technology Innovation

For Korea, the move from IS-95 to CDMA2000 would be a continuous innovation, both in terms of technology and business organization and structure. With its capabilities in CDMA technology and the help of Qualcomm, the move to WCDMA would also be a continuous technology innovation. However, it would be a slight disruption in business organization and structure to morph the purely IS-95 Korean mobile market to one that operated WCDMA.

The Position of an Organization in the Market

Within the local Korean market, the Korean government had the power to a set the 3G standard. It also had the right to allocate frequency spectra for 3G mobile commercialization and the influence to shepherd the Korean companies to pursue the industrial policies it chose to set.

With respect to the large markets of the U.S., Europe, and Japan, the Korean government had no power or influence regarding their 3G direction. With respect to the smaller markets in the rest of the world, for example, those in Southeast Asia, the Korean government could possibly influence their 2G and 3G development through political relations and financial deals. And for the two huge potential markets of China and India, which were growing economically but still lacked technological wherewithal comparable to Korea, the Korean government should be able to help the Korean companies, both the handset and the system vendors, to participate significantly there.

The Availability of Alternative or Substitutable Technologies

Korea was of course free to set its own 3G standard for the Korean market, or to adopt either the CDMA2000 or WCDMA standard, or to deploy some mix of these three. Outside of Korea, it was unlikely that any other market would choose to adopt an isolated Korean 3G standard given the line-up of WCDMA and CDMA2000. The question was really what choice Korea could make regarding CDMA2000 and WCDMA that would best advance its prospects globally.

Ownership and Deployment of IP (Intellectual Property)

For Korea's 2G development, Qualcomm had by and large supplied Korea with the full suite of CDMA technologies. For example, without Qualcomm's IC, the Korean companies would not have been able to build handsets or base stations. Qualcomm had been charging the Koreans 5.25% of the product Average Selling Price (ASP) for domestic sales and 5.75% for exports under a Most Favorable Royalty Rate (MFRR) (Mock, 2005). As the Korean mobile business grew, the Korean companies paid Qualcomm hundreds of millions of dollars. Although the exact number is not known, it is estimated that the Koreans had to pay the GSM IPR holders a total of 10% to 20% of handset price for the right to manufacture GSM handsets (YonHap News, 2005).

Although the Koreans had started to develop certain technologies on their own, and started to file patents, they had very few of the essential IPRs on CDMA2000 or WCDMA, mainly because they were late to the game of staking 3G IPR claims. The Koreans would have better opportunities to develop and own IPR on mobile broadband applications such as TV over handsets, even 4G technologies, because these were still virgin territories.

3.3.3. Korean Government's Interpretation of its Situation (CDMA2000)

Overall, the Korean government saw the development of next generation mobile as a significant new opportunity to further develop its communications and electronics capabilities and markets, thereby helping Korea become a world-class nation.

After its initial enthusiasm to develop its own 3G technology and standard, by 1998, the Korean government realized that its position in the global arenas (politics, market, and technology) was not strong enough for it to play a leadership role in defining the contours of 3G mobile technology.

The competition in 2G evolved to feature the upstart Qualcomm (with its disruptive CDMA innovation) challenging the established actors (with their TDMA technology). This competition was often seen as American technology against European technology, sometimes even portrayed as a holy war between CDMA and GSM. Although Korea benefited very much from its early adoption of CDMA technology, the Korean government saw no reasons to take sides, even if there were a 3G holy war between CDMA2000 and WCDMA. Its aim was to create benefits for the Korean nation from 3G development, regardless of which versions of 3G were to prevail.

Knowing CDMA technologies, and knowing how the CDMA2000 and WCDMA technologies were being developed, the Korean government understood that the CDMA2000 specification was much more technologically secure and ready for commercialization than the largely politically specified WCDMA. To be at the leading edge of developing 3G markets and equipment, the Korean government saw the necessity of commercializing CDMA2000 in Korea. Therefore, Korea had to remain on good terms with Qualcomm and support CDMA2000. In any event, Korea wanted to retain Qualcomm as an IC supplier when the time for attacking the WCDMA market came.

In the meantime, the Korean government could just as well let Qualcomm take the lead and bear the cost of championing CDMA2000 and cracking the heavily defended WCDMA market. Moreover, since Qualcomm would need the Korean mobile companies as partners in its CDMA2000 value network and as allies in penetrating future WCDMA markets, the Korean government saw the opportunity to bargain for a better IPR deal with Qualcomm. Perhaps Korea could even get indirect licensed use of GSM and WCDMA IPR through Qualcomm's cross-licensing deals with Nokia and others.

In hindsight, it would appear that the Korean government was also skillful in the art of maneuver warfare.

3.3.4. Korean Government's Strategies (CDMA2000)

Configuration of Value Network

The Korean government intended Korea to be a major global player in both the emerging CDMA2000 and WCDMA markets. It therefore aimed to guide the development of end-to-end value chain capabilities for both CDMA2000 and WCDMA. This included 3G service delivery, 3G software and content development, 3G network planning and design, and of course systems and handset design and manufacturing.

To this end, in 1999 the Korean government announced its plan to adopt both CDMA2000 and WCDMA for the Korean market and granted 3G licenses to three parties in 2000 to develop and implement the two 3G mobile technologies.

The Korean government did not want the existing mobile service providers to benefit disproportionately from implementing 3G. It set the policy that the three parties responsible for implementing 3G be formed as consortia composed of mobile service providers and other venture companies such as system developers or content providers. In this way, the Korean 3G value network would be more inclusive and the benefits from 3G could spread and help the growth of newer industries such as multimedia content development.

SK Telecom and KTF, the first and second largest mobile service providers in Korea respectively, founded independent consortia companies, SKIMT and KTICOM, with other venture companies to implement WCMDA. Of course, SK Telecom and KTF became the largest shareholders in these newly created companies. The Korean government also allowed LG Telecom, the third largest Korean mobile communications service provider, to lead a consortium to implement CDMA2000. Then, the Korean government assigned the necessary bands between the frequencies of 2 GHz and 3 GHz to the three consortia for implementing WCDMA and CDMA2000.

The Korean government's strategy was to strive to be on the cutting edge of developing 3G value networks, all the while promoting competences in all aspects of mobile broadband multimedia. The Korean government figured that, after building their 3G competences in Korea, Korean companies in the various areas of 3G mobile would proactively and opportunistically project themselves into value networks emerging outside Korea.

Formation of Standard-Setting

Because it saw Korea as still relatively weak in the global mobile market, the Korean government decided to cooperate with other standard-setting organizations in standardizing technologies.

First, the Korean government used TTA, Korea's voluntary standard-setting organization, to harmonize and consolidate the interests of the many commucations actors in Korea. With that accomplished, TTA then actively involved itself in the various international standard-setting organizations, such as ITU, 3GPP and 3GPP-2, to represent Korea's interests.

The consortium 3GPP-2 was found in 1998 to promote and coordinate research for CDMA2000.

It was set up as a counterpart to 3GPP which was formed for promoting WCDMA. Although 3GPP and 3GPP-2 were not strictly standard-setting organizations, they were the bodies through which future versions of WCDMA and CDMA2000, respectively, were to be researched, defined, and specified. Although 3GPP and 3GPP-2 were competing consortia, the Korean government participated in both to gauge and seize opportunities to interject Korean interests in the unfolding of the competing 3G technologies and markets.

Openness of IPR (Intellectual Property Rights)

Since IPRs were strategically important competitive weapons in the mobile industry, the Korean government strongly encouraged the Korean companies and research institutes to actively generate IPR from the technologies they were developing. IPR in mobile broadband multimedia technologies and applications were particularly promising for putting Korea in a better position to negotiate with non-Korean IPR holders for cross-licensing. Advanced research in broadband was also promoted for the possibility of gaining essential IPR in 3G applications, as well as in 4G.

With respect to CDMA technology, Korea was by and large still beholden to Qualcomm. Through tough negotiations, the Korean government was able to secure "most favored" licensing status from Qualcomm, which guaranteed that Korean companies would pay the least among Qualcomm licensees. On behalf of its IC customers, the most important of which were the Koreans, Qualcomm was able to insist on "pass through rights" on the cross-licensing part of its negotiation with WCDMA IPR holders. This meant that Korean manufacturers would possibly have low cost access to WCDMA IPR to produce WCDMA equipment, as long as they used Qualcomm's WCDMA ICs. However, "pass through rights" is still controversial issue.

3.4. The Emergence of the CDMA2000 Standard

Relevant information about the involvement of Qualcomm and the Korean government in the emergence of the CDMA2000 standard was discussed in the sections of QUALCOMM (CDMA2000) and The Korean Government (CDMA2000). Thus, this section will briefly describe the emergence of the CDMA2000 standard overall.

There was no doubt that Qualcomm's technologies dominated CDMA2000, and that it would be relatively easy for Qualcomm to persuade the mobile service providers who had adopted the IS-95 standard to move toward CDMA2000 technology because of the natural migration. This process could be accomplished without going through standardization by ITU, but recognition of CDMA2000 as a standard by international standard-setting organizations would better legitimize this technology and thus attract more actors.

Qualcomm formed an alliance with other system manufacturers (Motorola, Lucent and Nortel) in 1997 to assure mobile service providers that there would be multiple system suppliers and therefore they should not worry about being locked into a few suppliers. This alliance was intended to support CDMA2000 in becoming an international standard (Bekkers, 2001). Thus, Qualcomm submitted proposals to ITU through TIA.

Meanwhile, the Korean government wanted to further grow its mobile communications industry and market without being so technologically dependent on Qualcomm. For this reason, it formed a value network with other Korean actors including research institutes and manufacturers to develop 3G technologies and submit them as proposals to ITU through TTA.

The technology proposals that the Korean government and Qualcomm pursued for 3G standards were based on existing CDMA systems, so there was not much difference between their proposals in terms of technological characteristics (which relates to the aspects of the availability of complementary products or compatibility of products/services, the type of technology innovation, and the availability of alternative or substitutable technologies).

The difference between the Korean government and Qualcomm lay primarily in their positions in the global mobile communications market. Qualcomm, as an IPR holder and chip-maker, held most of the essential patents for CDMA technology and had strong capabilities in manufacturing CDMA-based chipsets, all of which the Korean government lacked. Qualcomm had gained a strong position in the global IS-95 market, while the Korean government only had an influential position in the domestic Korean market. Qualcomm also had a competent patents portfolio for CDMA but the Korean government did not. However, the Korean government controlled the most advanced CDMA market in the world and influenced very competent Korean CDMA systems manufacturers. Consequently, the Korean market was the largest market for Qualcomm and the Korean manufacturers were the company's largest buyers, so the Korean government's decision for a 3G mobile communications technology standard had significant implications for Qualcomm.

Qualcomm's interpretation was based on two values for its frontal and oblique moves. First, it valued CDMA as the most potential technology for a 3G standard. Second, it perceived that one way to expand its market and not stagnate as a minor player in the 2G arena was by pushing the global mobile communications market to move toward the 3G technology. In this way, Qualcomm would have the opportunity to penetrate the sealed GSM market by persuading GSM operators to adopt CMDA2000 when the industry moved to 3G.

The Korean government interpreted its situation based on its desire to boost its mobile industry and market continuously. The Korean government hoped to standardize its technology proposals that would create a more advantageous position for its national manufacturers. However, the Korean government was also willing to adopt other standards if they were more likely to be accepted in other global markets, because the early adoption of a potential technology standard would provide opportunities for the Korean manufacturers to gain necessary capabilities in manufacturing systems to export. In particular, due to the dominance of the GSM market, the Korean government was willing to adopt whatever technology that the GSM camp would implement, so the Korean manufacturers would be able to use the Korean domestic market to test their 3G systems.

In short, Qualcomm and the Korean government both wanted to push the global mobile communications market to 3G mobile technology in order to gain the opportunity to improve their positions through 3G standard-setting processes. Both preferred a 3G standard based on IS-95, so they would have an advantage in developing 3G technology systems by utilizing their existing resources and capabilities. However, the Korean government was also open to adopting other potential standards.

When the Korean government realized how difficult it was to persuade non-Korean actors to support its proposals in 1999, it dismissed its own proposals and decided to adopt the two technologies with the most potential – WCDMA and CDMA2000 – for its domestic market. The

Korean government believed that early adaptation and implementation of WCDMA and CDMA2000 in its domestic market would provide the Korean manufacturers advantages in developing and testing the systems.

4. EMERGENCE OF THE 3G STANDARDS CONFIGURATION

ITU had ambitions to set a single standard for the 3G global mobile market and planned to decide on this standard at the meeting in March 1999, but the meeting ended fruitlessly without any final decisions because of the conflicting interests among members and the on-going litigation between Ericsson and Qualcomm at the time (Bekkers, 2001) (see Figure 6).

Many mobile service providers, unhappy with this result, established Operators Harmonization Group (OHG) to solve this problem in 1999. Having multiple standards would be greatly disadvantageous to mobile service providers in terms of the incompatibilities between different standards and the large costs and long timeframe to implement 3G standards, so OHG took over the role of ITU and recommended four modes, shown in Table 3. Although mobile service providers wanted to have a single 3G mobile communica-

Figure 6. The timeline of development of the 3G standards

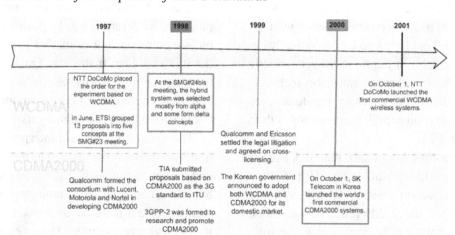

Table 3. OHG defined modes and requested changes for the 3G technology standards

Mode	Supported by	Changes
Direct Sequence WCDMA	NTT DoCoMo and many European actors who participated in NTT DoCoMo's experiment.	Chip rate down from 4.096 to 3.84 Mchip/s. Support ANSI IS-41 signaling in a second phase.
Multicarrier CDMA (CDMA2000)	Qualcomm and its allies.	Support for GSM/MAP signaling in a second phase.
TDD WCDMA	Siemens, its alliances and the Chinese government.	Harmonization with the Chinese TD-SCDMA.
EDGE/UWC-136	Some actors who wanted to continuously upgrade their systems based on existing D-AMPS or one of its revisions systems.	

(Source: Bekkers, 2001, p. 489)

tions technology standard, they realized that this would be impossible due to the conflicting interests among various actors such as systems manufacturers, governments, etc. Thus, they focused on harmonizing different standards by requiring changes in their specifications.

Figure 7 illustrates the migration paths among IS-95 (CDMA), GSM, D-AMPS, and PDC. The major manufacturers who had standardized WCDMA technology in Europe kept pursuing WCDMA as a global standard, while most actors who had supported IS-95 advocated CDMA2000. Some actors who had adopted D-AMPS insisted on following its continuous evolution route, EDGE, but other D-AMPS actors started to switch to WCDMA or CDMA2000. Time Division / Synchronous Code Division Multiple Access (TD-SCDMA), or TDD WCDMA, was another 3G technology that had been developed by the Chinese government and Siemens. Since this

Figure 7. The migration path of 2G to 3G standards (Source: CDMA Development Group Website [http://www.cdg.org] and Evolution of standards for 3G mobile communication, "W-CDMA v. CDMA2000," SIEMENS AG 2002)

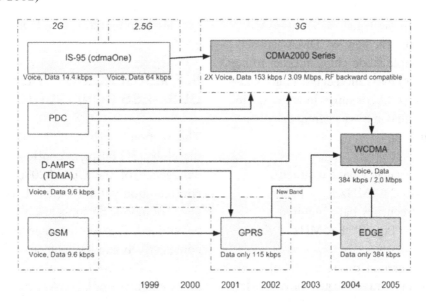

technology was not commercialized even by early 2008, it is not included in Figure 7. This technology mode and the relationship between the Chinese government and Siemens will be discussed more detail in the section of The Rise of Mobile Communications in China and India.

One thing to note in Figure 7 is the existence of 2.5G, between 2G and 3G. 2.5G technology such as General Packet Radio Service (GPRS) became a stepping stone to 3G when 3G technologies were not available, so mobile service providers could provide simple data delivery services. This was more applicable for GSM mobile service providers, because IS-95 (cdma-One) had the capacity to provide 2.5G services as well as 2G. The distinction between 2G, 2.5G and 3G is debatable for CDMA technology, because its natural migration path and the nature of CDMA means that 2G and 2.5G could be seen as overlapping, or 2.5G and 3G.

As shown in Figure 7, the proposed EDGE mode was used as another step to WCDMA.

The migration from IS-95 to the first sequence of CDMA2000 series requires minor modifications to base stations and handsets (see Figure 1). However, the migration to WCDMA from any technology requires new bandwidth and large investments, including new handsets, because WCDMA-only handsets do not have backward compatibility with the legacy GSM, D-AMPS, and GPRS systems. The new WCDMA handsets require dual or triple mode chips in order to be compatible. Battery life is another issue for WCDMA, because WCDMA consumes more energy in order to find a signal in a wider frequency band (UMTS Website).

- Minimum frequency band required (WCDMA) – 2 * 5MHz
- Minimum frequency band required (CDMA2000 1x) – 2* 1.25MHz

Regardless migration costs, it was obvious that the major actors who advocated for WCDMA tried to push not only the GSM market but also the PDC and A-TDMA markets to adopt WCDMA, while the CDMA camp also tried to attract these same markets to adopt the CDMA2000 standard. Through such back and forth disputes and reconciliations among actors, they settled on the 3G standards configuration by 2000 as below:

1. Agreed on three incompatible 3G standards – WCDMA, CDMA2000, TD-SCDMA (considering EDGE as a stepping standard to WCDMA).
2. Agreed on frequency bands, for example, 1920-1980 MHz paired with 2110-2170 MHz for WCDMA.
3. Accepted and defined different migration paths (see Figure 7).
4. Agreed on certain methods and protocols to connect between incompatible networks, for instance, dual- or triple-mode chipsets.

All in all, the 3G standard configuration emerged as a truce between the two intransigent camps of WCDMA and CDMA2000, with China allowed to pursue its dream of controlling its own 3G. In any case, the most important fact is that the global 3G standard configuration was set, at least for that moment. All the actors then moved to pursue their business strategies to dominate this business configuration (see Figure 8).

5. EMERGENCE OF THE GLOBAL 3G BUSINESS CONFIGURATION

This section briefly describes the emergence of the global 3G business configuration. First, the emergence of the CDMA2000 value network configuration will be sketched. Then, the emergence of mobile business in China and India will be outlined. This sets the stage for discussing the comparatively late development of WCDMA business, and how WCDMA actors tried to respond to the challenge of CDMA2000.

Figure 8. The configuration of 3G standards in mid-2008. Note: "WCDMA and CDMA2000" means that both standards have been introduced in the same national market. This picture is simplified and does not show standards adopted by every country. (Source: CDG, 2008; UMTS Forum, 2008)

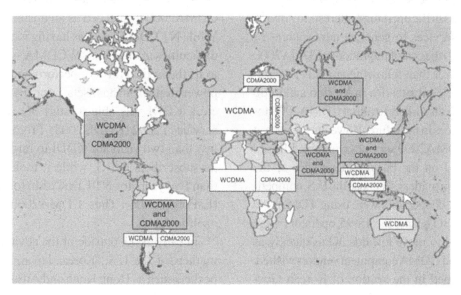

5.1. Emergence of the CDMA2000 Business Configuration

In early 1999, Qualcomm and Ericsson settled their CDMA patent litigation. Ericsson agreed to recognize Qualcomm's ownership of essential IPR for CDMA technology, worked out certain cross-licensing terms with Qualcomm on other IPR, and agreed to purchase Qualcomm's network division. This event signaled that Qualcomm would be legally able to pursue its business model for WCDMA, focusing on IP licensing and IC supply as outlined in the section of QUALCOMM (CDMA2000). Moreover, by selling its network division to Ericsson, Qualcomm hoped that Ericsson's European background would influence others to join the other system manufacturers (Motorola, Lucent and Nortel) in supporting CDMA2000.

Qualcomm recognized that sooner or later it would needed cooperation from major European actors if it was to promote CDMA2000 in Europe, or to successfully participate in the WCDMA market. However, it appeared that the European

actors were determined to block Qualcomm from cracking the European market. On the marketing and public relations side, the European vendors spread fear, uncertainty, and doubt about Qualcomm and CDMA2000, and portrayed the competition between CDMA2000 and WCDMA as a holy war (Mock, 2005). On the political side, the EU had a policy of "urging" its member countries to grant 3G licenses only to operators who selected 3G technologies compatible with other operators – this meant that in effect, the EU would not allow CDMA2000 into Europe, contrary to its proclaimed support of WTO principles, (Bekkers, 2001). In spite of the U.S. government's complaint against such anti-competitive EU policies, it appeared quite clear that CDMA will be locked out, effectively if not officially, from Europe (The EU did later change its policy. Officially, European license-granted operators would be free to adopt any technology standard that was a member of the IMT-2000 family as long as one of the granted operators per member country would implement the standard that ETSI recommended [Bekkers, 2001].).

In the U.S., the mobile service providers that had adopted the 2G IS-95 standard would naturally migrate towards 3G CDMA2000 technology, because (1) it was in the operators' interests to provide multimedia services to generate new sources of revenue and (2) the cost of migration to CDMA2000 was much lower than following any other migration path. Their subsidiaries in Mexico and Latin America were also likely to adopt CDMA2000 for their 3G services in the future. However, the pace at which 3G CDMA2000 services could be diffused would depend on the business strategies of the IS-95 service providers. Because the U.S. competitors of the IS-95 operators were using TDMA and had no clear 3G plans, the IS-95 operators were not compelled to move towards 3G as quickly as Qualcomm and CDMA equipment vendors wished.

As described in the section of Korean Government's Strategies (CDMA2000), the Korean government chose to implement both CDMA2000 and WCDMA in the Korean market. At the time, this decision appeared to be a significant blow to the future prospects of CDMA2000. But in reality, as to be explained presently, things turned out fine for Qualcomm and CDMA2000.

Although the Korean government designated two Korean parties (out of three) to implement WCDMA, it did not prohibit any of the Korean mobile service providers from migrating their existing IS-95 systems to CDMA2000, as long as they promised to eventually implement WCDMA according to a certain adjustable schedule. SK Telecom quickly launched the first commercial CDMA2000 system in the world on October 1, 2000. Some critics, particularly those who favored WCDMA, argued that the SK Telecom system, based on CDMA2000 1x technology, was only a 2.5G rather than a true 3G service. Nevertheless, SK Telecom's CDMA2000 voice and data service was very well received. The other Korean service providers, to protect their revenues and market, also rolled out CDMA2000 services quickly. Thus, in effect, the whole Korean mobile market had migrated to CDMA2000 by 2001.

In Japan, NTT DoCoMo was set on developing its WCDMA 3G service. Its only major competitor, KDDI, would have to move quickly from IS-95 to CDMA2000 to succeed as an upstart challenger. While NTT DoCoMo was having many technical difficulties in making its WCDMA system live up to its theoretical level of performance (excessive power-drain from handsets, poor signal handoffs, etc.), KDDI was able to start 3G (CDMA 1x) service in 2002 (KDDI, 2008). From then, it took less than two years for KDDI to migrate most of its subscribers to 3G, while it took a little less than five years for NTT DoCoMo to do the same (LaForge, 2006). Thus, CDMA2000 established itself in Japan.

In early 2000, outside of the advanced mobile markets of the U.S., Korea, Japan, Europe, and pockets such as Hong Kong and Australia, the rest of the world was not quite ready for 3G services. In fact, a large number of poor countries had very inadequate communications systems and no mobile services for the masses. Given this fact, Qualcomm and its partners pursued a three-pronged strategy: (1) promoting the GSM 1x technology as a viable way for existing GSM operators to migrate relatively painlessly to 3G; (2) continuing to try to persuade operators to adopt IS-95 for 2G service then move to CDMA2000 as the operators' business matured; and (3) using CDMA Wireless Local Loop technology (which is a fixed-central-base and limited-mobility-handset setup) to enable relatively poor countries to start mobile communications then move to CDMA2000 as operators' business matured.

Whether the targeted customer operators would decide to go for CDMA depended on many factors: (a) their business strategies; (b) their plans to provide future 3G services – if one were only concerned with 2G, then GSM would be a cheaper and safer choice than IS-95; (c) financial and other incentives offered by the vendors; (d) the political environment, particularly the operator countries' relations with Europe, the U.S., Korea and Japan.

By and large, for many reasons, the path of GSM 1x failed to attract anyone.

By structuring a special manufacturing licensing deal with the Chinese government, Qualcomm was able to obtain permission for its Chinese partner China Unicom to start IS-95 service to compete against the entrenched Chinese GSM operators. However, since China had its own ambitions for 3G, it was not clear how far CDMA2000 would proceed, if at all, in China (More on this in the next section).

Qualcomm was also able to help establish IS-95 operators in Indonesia, Malaysia, Pakistan, New Zealand, and other countries (CDG, 2008).

A significant success for Qualcomm was entering the huge potential market of India. India's major conglomerate Reliance Communications negotiated favorable terms with Qualcomm, and proceeded to introduce CDMA WLL service in India in 2003 (Business Wire, 2002; CDG, 2008). The CDMA WLL technology enabled Reliance to charge very low tariffs (as low as 50 rupees a month) and supply handsets to subscribers cheaply (Manish, 2004). The Korean vendors, particularly LG, were instrumental in setting up the WLL network and supplying the inexpensive handsets (India Telecom, 2003). This form of cheap mobile service was enthusiastically received by the Indian masses, and Reliance quickly built a nationwide service.

All in all, by 2004, the CDMA camp had established a nontrivial global footprint for 2G services, in preparation for CDMA2000. And CDMA2000 had been successfully commercialized as 3G mobile services in the core markets of Korea, the U.S., and Japan. In the meantime, GSM operators were still plodding along with their migration to interim 2.5G data service via GPRS, and having difficulties making WCDMA ready for full-scale commercialization.

But the fact remained that the majority of the world's operators (70% or more), were not going to adopt CDMA2000 for one reason or another. However, CDMA2000 had been established as the first mover in 3G, and thereby accomplished two strategic aims. First, in markets that allowed head

to head competition, it showed that CDMA2000 was indeed a better technology than WCDMA. Second, enabling CDMA2000 operators to generate significant ARPU from 3G services exerted pressure on WCDMA operators to speed up their move to 3G.

The story of how this early success helped Qualcomm and the Korean manufacturers to crack the WCDMA market will be told in the section of The Participation of QUALCOMM and Korea in WCDMA.

5.2. The Rise of Mobile Communications in China and India

The rapid industrial development of China and India over the last two decades have made the Chinese and Indian markets more important than ever in the world economy because they have the first largest (1.32 billion) and second (1.13 billion) largest populations in the world. This means that even if 30% of the Chinese population were to adopt cellular phones, that would be 396 million customers, larger than the entire U.S. population (304 million).

However, when the 1G market and the early 2G market in developed countries were growing in the late 1980s and early 1990s, China and India were not attractive markets because most of their population could not afford to have a handset or even a landline phone. For example, in China, the number of telephone lines per 100 inhabitants was 3.3 and mobile subscribers per 100 inhabitants was 0.29 in 1995 (Yan, n.d.).

In 1987, China offered its first mobile communications service based on Ericsson's TACS system in Guangdong (Yan, n.d.; Yu, 2005). Seven years later (1995), India introduced its first commercial mobile communications service (Shetty, 2007). Due to their late industrial development, their markets and manufacturers were technology followers and adopters rather than leaders and developers in many industries including mobile communications. Although China and India have risen to become the first and second largest mobile communications markets in the world, they had not played significant roles during the 1G and 2G technology standardizations. They just adopted available communications technology standards from foreign vendors. However, their choices for technology standards were significant for the many actors wanting to penetrate the Chinese and Indian markets. In the mobile communications industry, their choices for 2G technology standards in the mid-1990s became important. This was when more companies began to realize the potential of the mobile communications markets in China and India.

Both countries selected GSM for their 2G standard because there was no data to evaluate the commercial performance of IS-95 (CDMA) at the time. However, IS-95 was introduced later in China and India by new mobile service providers, China Unicom and Reliance India Mobile respectively.

Both countries were similar in terms of having growing large markets, but the evolution of their mobile communications industries was radically different. The Chinese communications market was strictly controlled by state-owned service providers (e.g. China Mobile and China Unicom), while the Indian market was run by many private operators. For example, in India, each of the twenty-eight states had services offered by a minimum of four to five service providers (Shetty, 2007).

Due to these different industrial structures, the Indian mobile communications market has been shaped by competition among service providers. The government policy set by the Telecom Regulatory Authority of India (TRAI) also supports free competition. However, in China, the Chinese government has more actively engaged in many affairs in the communications industry, and thus the Chinese mobile market has been shaped by many government regulations.

In 2007, the Indian government announced that it would auction 3G spectrum without imposing any 3G standards (Malik, 2007), which meant that

mobile service providers were free to adopt any 3G standard based on their existing infrastructure. The Indian mobile service providers would thus adopt WCDMA or CDMA2000 based on their 2G infrastructures. As with 1G and 2G, the decisions of the Indian mobile service providers would not affect standard setting for 3G technology, but it would significantly affect the prosperity of mobile communications vendors.

However, the Chinese government had different aspirations. It wanted not only its mobile communications market to grow fast but also its national industry to take a leading position in the global industry instead of following. Thus, it decided to develop its own 3G mobile communications technology standard, TD-SCDMA, that was based on TDMA and CDMA.

It is arguable whether TD-SCDMA was the Chinese' own standard. To respond to this argument and analyze whether the Chinese government's strategy affected the global 3G standardization process, let us review the development of TD-SCDMA.

It was surprising to many that the Chinese government proposed TD-SCDMA to ITU in 1998, because it had selected GSM as a 2G standard for the Chinese market in 1995 and therefore the government and Chinese manufacturers did not have any knowledge and capabilities for CDMA. Then, where did they gain the necessary knowledge and capabilities to draw up the proposal?

When ITU failed to standardize 3G technologies, OHG later defined four modes, with changes to allow for harmonization. OHG asked for changes in TDD WCDMA supported by Siemens to harmonize with the Chinese TD-SCDMA. This was one piece of evidence that there was a relationship between the Chinese TD-SCDMA and Siemens' TD/CDMA. In fact, the Chinese TD-SCDMA was developed through great cooperation with Siemens (Lu, 2000). As Willie W. Lu, chair of 3G'2000 in San Francisco stated in his speech, "Siemens helped a lot both in finance and ITU support." It is difficult to establish who first approached whom, but it is clear that Siemens and the Chinese government decided to cooperate with each other, so they developed Siemens' original TD/CDMA proposal further to standardize TD-SCDMA as an international standard (Clark, 2007). Siemens, which had failed to standardize its technology in ETSI, wanted to globally standardize its technology, an aspiration that coincided with the aspiration of the Chinese government to turn its national manufacturers into technology innovators rather than followers.

The development of TD-SCDMA was led by the China Academy of Communication Technology (CATT) at the beginning. The CATT founded Datang Telecom, a Chinese communication equipment vendor, in 1998. Siemens has been heavily involved in developing TD-SCDMA systems with Datang since then (Yu, 2005).

If the Chinese government and Siemens had been able to standardize TD-SCDMA as early as NTT DoCoMo commercialized WCDMA in 2001, they could have influenced the industrial configuration of 3G standards as they wished. However, they had not commercially launched mobile communications service on TD-SCDMA infrastructure even by April 2008. By this time, the WCDMA and CDMA2000 systems had been established worldwide with more than 140 million and 70 million subscribers respectively (Clark, 2007).

The Chinese government has delayed licensing 3G, but announced that it will include WCDMA and CDMA2000 as well as TD-SCDMA in its 3G landscape (Morse, 2007). This means that there will be three standards in the Chinese 3G market, which decreases the strength of TD-SCDMA, because the Chinese mobile service providers will tend to choose the most cost-effective technology considering their existing infrastructures. From this perspective, TD-SCDMA would not be their favored technology, because TD-SCDMA still has technical and economic problems. For example, TD-SCDMA systems require larger antennas and more base stations than WCDMA and CDMA2000, which would make it more costly

than implementing already proven WCDMA and CDMA2000 systems (Clark, 2007).

Given this situation, it can be predicted that TD-SCDMA will not significantly affect the self-organized configuration of the 3G industry. Even if TD-SCDMA systems is commercially launched, it is likely to remain a local standard in China and not affect the current global configuration of 3G standards, because the global mobile communications market with the exception of China has already emerged with 3G services based on WCDMA or CDMA2000 and it will not switch to TD-SCDMA.

5.3. The Participation of Qualcomm and Korea in WCDMA

As mentioned, NTT DoCoMo started to provide the world's first WCDMA service, the so-called Freedom of Mobile Multimedia Access (FOMA), in 2001. Before launching the service, NTT DoCoMo experienced several technical difficulties, so it postponed the commercial launch date a few times. However, its difficulties continued even after the commercial debut. It was widely acknowledged that "FOMA was a failure until late 2003" (Adamson, 2004), because NTT DoCoMo did not attract as many 3G customers as it had expected and instead lost them to KDDI. For example, by August 2003, there were 9.7 million subscribers with 3G CDMA2000 phones from KDDI, while only 785,000 customers had subscribed to NTT DoCoMo's FOMA service, nearly two years after NTT DoCoMo's commercial 3G launch (Adamson, 2004). The key failure was technological problems such as backward incompatibility and the immaturity of the handsets (short battery lives and heavy weight), while KDDI had the backward compatibility between cdmaOne and CDMA2000, and could also provide a greater variety of savvy handsets.

Ironically, NTT DoCoMo was able to solve these problems with help from KDDI in late 2003 (Adamson, 2004). From October 2003 to

March 2004, the number of FOMA subscribers increased drastically by 300 percent (over three million) (Adamson, 2004). Considering the fact that Qualcomm not only persuaded KDDI to adopt CDMA but also worked closely in implementing and commercializing CDMA, it can be deduced that Qualcomm indirectly (or perhaps directly) helped NTT DoCoMo to clean up its technical problems. In light of this situation, it was not surprising that NTT DoCoMo and Qualcomm announced their cooperation in promoting global WCDMA deployment in 2004 (Nikkei Electronics Asia, 2004).

After Qualcomm paved this road to penetrating the Japanese WCDMA market, the Korean manufacturer LG and NTT DoCoMo agreed to develop a dual-mode 3G FOMA handset that would work on both WCDMA and GSM/GPRS networks based on Qualcomm's chipsets (Adamson, 2005).

In Europe, when the European actors agreed on the UMTS standard, they reconciled the hybrid specification (mostly from WCDMA and somewhat from ATDMA) without detailed information in their hurry to submit a unified European proposal to ITU, while forming 3GPP to define a more detailed technological specification. This made the technical design more complex and the migration path to the European version of WCDMA more difficult. Although all these technical problems could have delayed European migration to 3G, the European actors successfully fortressed their European market from the CDMA2000 standard.

However, there was another factor that caused the delay of 3G implementation in Europe. Many European governments started to sell necessary frequency bands to service providers. For example, the U.K. collected total 37.5 billion Euros and Germany gathered total 50.5 billion Euro by auctioning frequency bands to multiple companies, which put great financial pressure on mobile service providers. Basically, the companies did not have enough financial resources to install 3G systems after investing so much capital for 3G licenses and frequency bands.

Having spent so many resources and observing the difficulties that NTT DoCoMo went through in commercializing WCDMA, the European mobile service providers started to realize that the European vendors they relied on had pushed the UMTS standard with their own interests in mind, and that their governments had taken advantage of them to secure finances. Losing faith in their European manufacturers and their governments made the mobile service providers more market-driven. This dissent between mobile service providers and other actors opened up a weak point in their existing value network that allowed Qualcomm and the Korean manufacturers to come in to provide 3G systems for the European 3G mobile communication market.

It was true that the European actors and NTT DoCoMo defined WCDMA as the heir of the GSM market and blocked out CDMA2000 with their technology standards strategies. But there were still weak points that Qualcomm and the Korean actors could attack. Thus, they failed to block out Qualcomm and the Korean actors, because their standards and business strategies were not just about the holy war between WCDMA and CDMA2000. This consequence affected some mobile service providers such as NTT DoCoMo and Vodafone in their aspirations to be global mobile operators. For example, NTT DoCoMo has struggled not to lose its subscribers to KDDI in its domestic market. It is doubtful how much NTT DoCoMo can focus on international expansion when it is having a hard time in its domestic market.

5.4. Response of WCDMA to the Challenge of CDMA2000

The conventional idea of war between WCDMA and CDMA2000 is over. As shown in the previous section, the actors (especially, manufacturers) compete against each other in both standards markets employing very sophisticated business strategies. Of course, they will compete in the

TD-SCDMA market when it becomes commercialized and appears lucrative.

It was too late for the mobile service providers that implemented WCDMA to regret their choice and blame other actors. Thus, they actively tried to improve and increase the performance of their networks while decreasing costs. For example, they started to use High Speed Packet Access (HSPA), which is a set of mobile telephony protocols. The protocols in this set such as High-Speed Downlink Packet Access (HSDPA) and High-Speed Uplink Packet Access (HSUPA) can improve and extend the performance of WCDMA networks. In another example, mobile service providers have also tried to implement IP networks in the backend instead of traditional mobile or cellular networks to decrease costs by using Internet for voice and data delivery. The incorporation of Internet-related technologies in their networks is directly related to evolution to 4G and the convergence of networks between the communications and computing industries, as will be explained in the following chapter.

REFERENCES

Adamson, W. (2004). How NTT DoCoMo stumbled in the race to 3G and its amazing recovery. *imodestrategy.com*. Retrieved April 10, 2008 from http://www.imodestrategy.com/2004/06/how_ntt_docomo_.html

Adamson, W. (2005). DoCoMo taps LG for roaming revenue expansion. *imodestrategy.com*. Retrieved April 10, 2008 from http://www.imodestrategy.com/2005/06/docomo_taps_lg_.html

ARIB IMT-2000 Study Committee. (1998). *Japan's revised proposal for candidate radio transmission technology on IMT-2000: W-CDMA*. Japan: ARIB, Association of Radio Industries and Businesses.

Bekkers, R. (2001). *Mobile telecommunications standards: GSM, UMTS, TETRA, and ERMES*. Boston, MA: Artech House.

Boyd, J. (2008). *Observe-orient-decide-act*. Retrieved April 10, 2008 from http://www.d-n-i.net/dni/john-r-boyd/

Business Wire. (2002). QUALCOMM makes commitment for $200 million strategic investment in reliance communications limited, reliance plans to deploy nationwide CDMA network in India. *Business Wire*. Retrieved May 8, 2012, from http://findarticles.com/p/articles/mi_m0EIN/is_2002_Jan_10/ai_81561801

CDG. (2008). 3G CDMA worldwide diffusions. *CDMA Worldwide Database*. Retrieved April 10, 2008 from http://www.cdg.org/worldwide

Channing, I. (1998). Crunch time for UMTS. *GSM World Focus*, 51-55

Clark, D. (2007). China misdials on mobiles. *Far Eastern Economic Review, 170*(10), 52-57.

Communications Week International. (1997, June 30). Euro backing boost for Japan's W-CDMA. *Communications Week International*, 25.

ETSI. (1988). *Universal mobile telecommunications system (UMTS), concept groups for the definition of the UMTS terrestrial radio access (UTRA) (Report TR 101 397 v3.0.1)*. Sofia Antipolis, France: ETSI.

ETSI. (1996, July). *ETSI technical report 314*. Sofia Antipolis, France: ETSI.

ETSI. (1999). *ETSI SMG moves to the selection of a 3rd generation radio access system*. Sofia Antipolis, France: ETSI.

ETSI. (2008). *IPR online database*. Retrieved May 8, 2012, from http://Webapp.etsi.org/ipr/

Forum, U. M. T. S. (2008). UMTS commercial deployments. *Wireless Intelligence*. Retrieved May 8, 2012 from http://www.umts-forum.org/content/view/2000/98/

Garrard, G. A. (1998). *Cellular communications: Worldwide market development*. Boston, MA: Artech House.

GSM MoU Association. (1999). *GSM MoU association chairman calls on manufacturers to end technology battle*. GSM MoU Association.

Gupta, P. (2008). *EDGE! Will TDMA and GSM ever meet? Mobile wireless communications tomorrow*. Retrieved May 8, 2012, from http://www.wirelessdevnet.com/channels/wireless/training/mobilewirelesstomorrow5.html

Holma, H., & Toskala, A. (2000). *WCDMA for UMTS: Radio access for third generation mobile communications*. New York: Wiley.

India Telecom. (2003, January 1). LG gets $104 million order from reliance. *India Telecom.*

Kalavakunta, R., & Kripalani, A. (2005, January). Evolution of mobile broadband access technologies and services - Considerations and solutions for smooth migration from 2G to 3G networks. In *Proceedings of the IEEE International Conference on Personal Wireless Communications (ICPWC)*, (pp. 144-149). IEEE.

KDDI. (2008). *KDDI history*. Retrieved May 8, 2012, from http://www.kddi.com/english/corporate/kddi/history/index.html

LaForge, P. (2006). Race for the future: The 3G mobile migration. *Telephony, 247*(14), 2–8.

Lu, W. W. (2000). *China 3G: TD-SCDMA behind the Great Wall*. Paper presented at 3G'2000. San Francisco, CA.

Malik, O. (2007). India finally has a 3G plan: WiMax in the mix, too. *GIGaom*. Retrieved May 8, 2012, from http://gigaom.com/2007/11/13/india-finally-has-a-3g-plan-wimax-in-the-mix-too/

Manish, C. (2004). Minimize your expenses using reliance phones. *Reliance India Mobile Review*. Retrieved May 8, 2012, from http://www.mouthshut.com/review/Reliance_India_Mobile-55473-1.html

Mobile Communications International. (1996, November). Japan shakes up for 3rd generation. *Mobile Communications International*, 16.

Mobile Europe. (1997, July/August). European commission calls for swift response on UMTS. *Mobile Europe*, 15.

Mock, D. (2005). The qualcomm equation: How a fledgling telecom company forged a new path to big profits and market dominance. New York: AMACOM (American Management Association).

Morse, J. (2007, June 15). China steps up 3G tempo. *WirelessWeek.*

Nandhini, V. (2008). *Migration of GSM networks to GPRS: Wipro technologies*. Retrieved May 8, 2012, from http://www.tslab.ssvl.kth.se/csd/projects/0308/gsmtogprs.pdf

Nikkei Electronics Asia. (2004, November). Qualcomm, NTT DoCoMo to promote global WCDMA deployment. *Nikkei Electronics Asia*. Retrieved May 8, 2012, from http://techon.nikkeibp.co.jp/NEA/archive/200411/343877/

Noam, E. M. (1992). *Telecommunications in Europe*. New York: Oxford University Press.

Pilato, F. (2004, November 18). GSM takes 74% share of global wireless market. *Mobile Magazine*. Retrieved April 10, 2008 from http://www.mobilemag.com/content/100/104/C3425/

Shetty, R. (2007). The talkative Indian. *Far Eastern Economic Review, 170*(10), 57.

Steinbock, D. (2001). *The Nokia revolution: The story of an extraordinary company that transformed an industry.* New York: AMACOM.

Sun-Tzu. (1994). *The art of war* (Sawyer, R. D., Trans.). Boulder, CO: Westview Press.

Vojcic, B., Pickholtz, R., & Stojanovic, I. (1991, June). *A comparison of TDMA and CDMA in microcellular radio channels.* Paper presented at the International Conference on Communications. Denver, CO.

Westmand, R. (1999). The battle of standards and their road to peace. *On the New World of Communication, 1*(1), 28-28.

Yan, X. (n.d.). *3G mobile policy: The case of China and Hong Kong, China.* International Telecommunication Union.

YonHap News. (2005, January 3). GSM royalty, 20% of handset price, need a plan to cope. *YonHap News.*

Yu, J. (2005). The national champion in ICT standard competition: Evidence from an emerging country. In *Proceedings of IEEE Conference,* (pp. 114-120). IEEE.

Chapter 8

Beyond the 3G (Third Generation) of Mobile Communications Technology Standards

EXECUTIVE SUMMARY

How the decade-long unfolding of 3G is being morphed into the convergence of communications and computing is described. With this new technology direction, actors in both the mobile and computing industries have started fighting to define the next generation mobile standard.

1. FROM 3G TO 4G AND THE CONVERGENCE OF NETWORKS BETWEEN THE COMMUNICATIONS AND COMPUTING INDUSTRIES

As the unfolding of 3G business showed, the global mobile communications industry has become more complex than ever because of the interaction of technologies, government policies, standards choices, and the range of the market from low-end to high-end products/services. During the 1G and 2G periods, actors competed to

DOI: 10.4018/978-1-4666-4074-0.ch008

define a standard in rather homogeneous regional markets. Then, they competed for market share within the standard once a standard was defined. In the 3G arena, actors had to not only compete to define standards but also compete within a standard and over standards in the same market in heterogeneous market conditions in various regions around the world.

Actors needed different strategies for regions where standards were committed and where standards were not committed. Even regions that had committed standards (such as Europe and Japan) could be divided into two types of regions. One had a committed single standard (e.g. Europe), and the other opened free competition between standards (e.g. Japan). In both types of regions, the issues that actors had to deal with were (1) timing of commercialization of a standard and (2) performance of a committed standard. These were especially important for actors in regions of free competition. If one standard were commercialized earlier than the other, it would provide first mover advantages. When two standards competed against each other in the same market, the standard with better performance would be more likely to win over the other if all other conditions were the same. For the market with a committed single standard, the actors still considered these two issues even though their standard did not have to compete against others in their market. The launch timing and performance of their selected standard would compete against other standards for the regions that had not committed to any standard. Thus, the actors that wanted to persuade other actors in other regions to adopt their standard still had to consider these issues in order to provide a better evaluation on their standard than other competing standards.

In the regions that had not committed to any 3G standard, the issues were the choice of (1) a standard and (2) system vendors. Each camp tried to persuade mobile service providers and governments in undecided regions to select their standard, so major system vendors could gain advantages in exporting their products.

These different types of markets co-existed, so actors in the mobile industry needed to develop their business to deal with different situations around the world simultaneously. As shown in the CDMA2000 case, many actors do not want to fight against each other just for sake of a standard. As the Korean government did, many actors look for their own benefits and split their bets on more than one potential standards if possible. Moreover, they are willing to switch one standard to another standard that is more potential to have a larger market.

The ongoing evolution of the global mobile communications market is becoming more complex as it moves toward 4G technologies. After the Internet revolution in the late 1990s, people have wanted to be connected to the Internet at all times. Thus, the data delivery affected the evolution of the 3G mobile market, all networks (e.g. wired, mobile, and satellite) are converging to IP-enabled network, so people can be connected to Internet regardless the different network types. This change in the environment opens up new opportunities for actors in the communications industry, computer industry, and broadcasting industry.

For example, the mobile communications market was separate from the computer industry even in the early 3G period. People used computers (including desktop personal and laptop computers) for connecting to the Internet, while carrying cellular phones for making and receiving phone calls and delivering simple data (e.g., SMS). However, as the Internet network has expanded and other technologies such as Voice-over-IP have developed, people can make phone calls through the Internet using their computers without cellular or landline phones. They can also connect to the Internet through their cellular phones or other variant of portable handsets (e.g., PDA and tablet computer) without desktop and laptop computers. This means that actors in the computer industry such as Internet service providers and portal service providers can offer services that only mobile service providers used to offer and vice versa.

All necessary hardware and software are in the process of being developed to support this convergence. During the convergence process, more technology standards are required to make these different systems compatible and interoperable to communicate with each other seamlessly. Actors who have acknowledged the significance of technology standards through the evolution of the ICT industry have tried their best to maneuver their standards strategies along with their business strategies. For example, Intel, which had long tried to be part of the mobile industry but failed, and its allies (e.g. Motorola and Samsung) were standardizing Worldwide Interoperability for Microwave Access (WiMax) as a next general standard of

WCDMA and CDMA2000. Meanwhile, one of the enthusiastic advocators of CDMA2000, Qualcomm proposed Ultra Mobile Broadband (UMB) for the 4G standard through the 3rd Generation Partnership Project 2 (3GPP2). From the WCDMA camp, that had successfully defended the largest mobile market since the 2G standard (GSM), also proposed its own next generation standard, Long Term Evolution (LTE) through the 3rd Generation Partnership Project (3GPP) to keep protecting their hegemony in the global mobile communications market.

More actors bring more competing standards into the overlapping market that has been created by the network convergence. For example, Cisco Systems, Inc. and Sonus Network, Inc. are providing mobile networking systems. Apple and Google also joined to provide handset devices and mobile operating systems. More actors from different industries imply more complex and dynamic interactions in the evolution of the overlapping industries – this is no longer about the mobile communications industry only. This chapter briefly explains how the 4G mobile communications technology standards have evolved.

2. TECHNOLOGICAL BACKGROUND

The main differences between 3G and 4G are capacity, data rate, security, and access technology to the Internet. It means that users can safely access to Internet from any kind of handset for large capacity applications such as video games and movies. Traditional circuit-switched telephony service will not be supported anymore over 4G systems. Moreover, various technologies such as Orthogonal Frequency Division Multiple Access (OFDMA) and multiple-input multiple-output (MIMO) communications replace the spread spectrum radio technology used in 3G systems. In this section, OFDMA and MINO are briefly explained.

2.1. Orthogonal Frequency Division Multiple Access (OFDMA)

Orthogonal Frequency Division Multiplexing (OFDM) is a technique to transmit large amounts of digital data on multiple carrier frequencies. The technology enables to split the radio signal into multiple smaller sub-signals in order to simultaneously transmit them over different frequencies. OFDMA was developed to move OFDM technology from a fixed-access wireless system to a cellular system with mobility. The technological principle is the same except that more flexibility was added. In OFDMA, sub carriers are grouped into larger units, so called sub channels, and these sub channels are further grouped into bursts that can be allocated to mobile users. The flexibility of burst allocation that can be changed from frame to frame allows the base station to dynamically adjust the bandwidth usage. As a result, the spectral efficiency is increased.

2.2. Multiple-Input and Multiple-Output (MIMO)

Multiple-Input and Multiple-Output (MIMO) as a wireless technology utilizes multiple transmitters and receivers to transfer more data at the same time. This technology consider a radio-wave phenomenon (called multipath) where transmitted information bounces off walls, ceilings, and other objects, and then reaches a receiving antenna multiple times through different angles at slightly different times. The MIMO technology leverages this phenomenon by using multiple, "smart" transmitters and receivers with an added "spatial" dimension to improve communication performance.

Based on this technological background, we move to each technology that was potential to be a 4G standard.

3. ULTRA MOBILE BROADBAND (UMB)

The UMB systems that Qualcomm proposed through 3GPP2 were developed based on Internet Protocol (IP) networking technologies and the goal of the systems was providing services with peak rates of up to 280 Mbit/s. This system had backward compatibility with existing CDMA2000 1X and 1xEV-DO systems. As mentioned, Qualcomm led developing and standardizing this system. Thus, Qualcomm's situation, interpretation of the situation, and strategies are reviewed.

3.1. Qualcomm's Situation

Organization's Capabilities to Meet Market Needs and Opportunities

Qualcomm as one of the global mobile chip-makers had capabilities to meet market needs and opportunities in term of developing mobile chipsets.

The Availability of Complementary Products or Compatibility of Products/Services

Qualcomm was able to make the UMB technology to be compatible with existing CDMA2000 1X and 1xEV-DO systems, but it definitely needed other companies to work with in order to provide a complete product or service.

The Type of Technology Innovation

There were new components (e.g., OFDMA and MIMO) on the 4G technologies in general, but the overall innovation level was considered continuous innovation.

The Position of the Organization in the Market

Qualcomm had a dominant position as a mobile chip-maker in the CDMA mobile phone market. Due to its almost monopoly position and the royalty fees it had charged, many companies especially from the WCDMA camp had been cautious about Qualcomm's movements.

The Availability of Alternative or Substitutable Technologies

There were two technologies, WiMax and LTE competing against UMB.

Ownership and Deployment of IP (Intellectual Property)

Qualcomm claimed that it had many patents related to the UMB technology.

3.2. Qualcomm's Interpretation of its Situation and its Strategic Direction

Although Qualcomm introduced the CDMA technology late in the 2G period, it successfully standardized IS-95 (cdmaOne). It also successfully developed a WCDMA chipset and gained decent market share. Qualcomm held an almost monopoly position on the CDMA mobile phone market, but it met strong competitions from Infineon, Texas Instruments and Broadcom in the 3G WCDMA market. It implied that Qualcomm gained the almost monopoly position when it was able to standardized its technology. This perspective made Qualcomm diligently try to standardize its technology in the 4G period as it did in the 3G period with the CDMA2000 technology.

3.3. Qualcomm's Strategies

Configuration of Value Network

Qualcomm definitely needed other companies to form a value network. Therefore, it was natural for Qualcomm to look for companies from their existing value networks of the CDMA2000 standard to build a value network for UMB.

Formation of Standard Setting

Along with Qualcomm's strategy on the configuration of value network, it tried to standardize the UMB technology through 3GPP2 that basically supported and advocated the IS-95 and CDMA2000 standards.

Openness of IPR (Intellectual Property Rights)

One of the most important revenue streams for Qualcomm came from royalty fees by licensing its patents. Thus, Qualcomm tried to have as many patents as possible related to the UMB technology. Then, Qualcomm were willing to license its patents and wanted to use them as bargain chips.

4. WIMAX (WORLDWIDE INTEROPERABILITY FOR MICROWAVE ACCESS)

WiMAX was developed and promoted by IEEE (Institute of Electrical and Electronics Engineers) and the WiMAX Forum. Especially, Intel as one of the largest investors and supporters put a lot of efforts to develop and standardize the technology along with the Korean actors such as the Korean government and Samsung. In Korea, a similar version and compatible with WiMAX, so-called

Wireless Broadband (WiBro) was under developing. Intel and Samsung agreed to guarantee the compatibility between WiMAX and WiBro in November 2004. WiBro uses 8.75/10 MHZ as a channel bandwidth, time-division duplexing and OFDMA for multiple accesses. Two Korean telecommunications operators (SKT and KT) began to provide the first commercial service over the WiBro network in June 2006. In this book, the term of WiMax includes WiBro as a part of IEEE 802.16e standard. Three important actors' (Intel, Samsung, and Korean government) situations, interpretations of the situations, and strategies are presented below.

4.1. The Situations of Intel, Samsung, and the Korean Government

Organization's Capabilities to Meet Market Needs and Opportunities

Intel was the most well-known chipmaker of microprocessors in the computer industry. It had wanted to participate in the mobile communications industry, but had not been successful. Intel was believed to have capabilities to develop a mobile chipset as it had capabilities to product Wi-Fi related products. WiFi as one of the most popular technologies in a computer network area, which makes an electronic device possible to exchange data wirelessly.

SamSung as one of the largest electronic conglomerates and mobile handset manufacturers had capabilities to develop a potential technology for a 4G mobile standard.

The Korean government had various research institutes (e.g., Telecommunications Technology Association of Korea [TTA]), so it had some capabilities to develop technologies by incorporating with the Korean national companies such as Samsung and LG through these research institutes.

The Availability of Complementary Products or Compatibility of Products/Services

WiMax series were evolved from a Wi-Fi and so-called the fixed WiMax (802.16d). Based on this history, it offered compatibility with most Internet devices.

The Type of Technology Innovation

Although WiMax had new components that allowed the mobility of a device (e.g., laptop) including OFDMA and MIMO, it was continuous innovation, because it was evolved from the existing technology.

The Position of the Organization in the Market

Intel had a strong market position in the computer (portable laptop and personal computer) industry, but it was a new entrant to the mobile communications industry. Meanwhile, Samsung had built a strong position as a mobile handset manufacturer in the global mobile communications market. The Korean government could influence on the Korean market by setting a standard, but not other markets.

The Availability of Alternative or Substitutable Technologies

There were two technologies, UMB and LTE competing against WiMax.

Ownership and Deployment of IP (Intellectual Property

Intel and Samsung were known to have some IPs relevant to WiMax.

4.2. Intel's, Samsung's, and the Korean Government's Interpretations of their Situations and their Strategic Directions

When a market is stable, it is difficult for new entrants to enter the market. However, when a market is moving toward a next generation technology standard that has not been defined, it opens a great opportunity for companies to enter the market. This was the situation for Intel to tap into the mobile communications market. Especially, Intel's technology capabilities could contribute in developing a next generation technology. Therefore, Intel foresighted and prepared itself to grasp this opportunity.

Although Samsung was one of the largest mobile handset manufacturers, it had relied on other companies (e.g., Qualcomm) for chipsets and other components. It was a good opportunity as well for Samsung to standardize a technology that it had more capabilities so that it could be more independent from other suppliers. Having more own technologies also meant less depending on patents from other companies. In this case, Samsung could decrease costs by diminishing royalty fees caused by getting licenses from IP owners.

The Korean government had witnessed a large portion of financial resource moving toward foreign companies because of royalty fees that the Korean manufacturers had to pay. To minimize this financial resource flowing to foreign companies, the Korean government had ambition to globally standardize its domestically developed technologies. The period that the mobile communications industry was seeking a 4G mobile technology standard opened this opportunity window for the Korean government to realize its ambition.

4.3. The Strategies of Intel, Samsung, and the Korean Government

Configuration of Value Network

As shown above, the interpretations of three actors' (Intel, Samsung, and the Korean government) situations were overlapped to agree on the same goal, standardizing the WiMax technology. Therefore, they could collaborate to build a value network together by agreeing on the compatibility between Intel's WiMax and the Korean WiBro.

Formation of Standard-Setting

All of the three actors were aware of that it would be very difficult to standardize a technology by themselves. Therefore, they tried to standardize it through Institute of Electrical and Electronics Engineers (IEEE) that wrote the fundamental technology specifications of WiFi and through the WiMAX Forum that advocated the WiMax technology.

Openness of IPR (Intellectual Property Rights)

Although Intel and Samsung had some IPs on the WiMax technology, it was their interests to diffuse the technology as much as they could before the commercialization of LTE to take first mover's advantage. Therefore, their strategies on IPs were providing an easy way for companies to access necessary licenses. In order to do that, they formed the Open Patent Alliance with four other firms (Cisco, Alcatel-Lucent, Clearwire and Sprint) to help other companies obtain access to necessary patents related to WiMax (Homeland Security News Wire, 2008).

5. LTE (LONG TERM EVOLUTION)

Long Term Evolution is considered the next generation of WCDMA suggested by the standard-setting organization of 3rd Generation Partnership Project (3GPP). It implies that the WCDMA camp was pushing LTE to maintain the WCDMA turf from other potential standards. LTE would provide backward compatibility with the GSM and WCDMA technologies and increase peak data rates with 100 Mbps downstream and 30 Mbps upstream, which could be improved in future. LTE would utilize the OFDMA and MIMO antenna technologies. As LTE was portrayed as the next generation of the WCDMA standard, it was not surprised to find out that the actors who had involved in the WCDMA development (e.g., Ericsson) actively participated in the LTE standardization. One thing different from the WCDMA case was that there was not a great split among actors who supported LTE. There were many actors who involved in developing and standardizing LTE, but their situations, interpretations, and strategies were similar to those in the case of WCDMA. This is why the situations of Ericsson and Nokia Siemens Networks (as representing companies), their interpretations of the situations, and their strategies are reviewed, even though NTT DoCoMo first proposed LTE.

5.1. The Situations of Ericsson and Nokia Siemens Networks

Organization's Capabilities to Meet Market Needs and Opportunities

Ericsson had successful experiences on developing technologies for standards such as GSM and WCDMA. Thus, there were no doubt that it had technological capabilities to meet market needs and opportunities. Nokia Siemens Networks had founded as a joint venture between Nokia and Siemens in 2007 to manufacture telecommunications equipments. Nokia Siemens Networks had also technological capabilities inherited from the two mother companies, Nokia and Siemens.

The Availability of Complementary Products or Compatibility of Products/Services

As mentioned, LTE was considered to be the next generation of WCDMA. Therefore, LTE was expected to have backward compatibility with WCDMA and the relevant series such as Enhanced Data rates for GSM Evolution (EDGE) and High Speed Packet Access (HSPA).

The Type of Technology Innovation

Although LTE had new components such as OFDMA and MIMO, it was continuous innovation.

The Position of the Organization in the Market

Ericsson had the world largest market share in the global mobile telecommunications equipment market. Nokia Siemens Networks were lately founded but were able to penetrate and establish their position in the mobile telecommunications equipment market.

The Availability of Alternative or Substitutable Technologies

There were two technologies, UMB and WiMax competing against LTE.

Ownership and Deployment of IP (Intellectual Property)

Ericsson claimed that it had 25% of essential patents on LTE (Fitchard, 2010). Nokia Siemens Networks also claimed that they had the most essential LTE patens. Although it was difficult to say how many essential patents they had were valid and legitimate, it was sure that they had some essential LTE patents.

5.2. Ericsson's and Nokia Siemens Networks' Interpretations of their Situations and their Strategic Directions

Ericsson and one of the mother companies of Nokia Siemens Networks, Nokia were able to protect and move the GSM market to the 3G period by standardizing the WCDMA technology that they supported. [Note: the unfolding tale that they met fierce competitions from other companies such as Samsung and LG in the WCDMA market, was another story after successfully standardizing their technology.] Although Qualcomm, Intel, and the Korean actors started earlier to develop and standardize a possible 4G standard, it was very significant for Ericsson and Nokia Siemens Networks to propose, develop, and standardize their technology for a 4G mobile communications standard in order to protect their positions and the WCDMA turf. Especially, considering the strong competitions from the Korean handset manufacturers in the WCDMA market, leaving those Korean actors to standardize WiMax as a 4G standard would be crisis for Ericsson and Nokia Siemens Networks. Their interpretation was moving the WCDMA market to the 4G period with the technology standard that they developed and supported.

5.3. The Strategies of Ericsson and Nokia Siemens Networks

Configuration of Value Network

As it was obvious that the LTE technology was portrayed as the next generation of the WCDMA, Ericsson and Nokia Siemens Networks tried to configure a value network with actors from the WCDMA camp including two mother companies, Nokia and Siemens as well as the WCDMA and LTE initiator, NTT DoCoMo. In addition, they wanted to persuade the Chinese actors to join their value network with another version of LTE, called TD-LTE. Considering the Siemens' involvement in developing TD-SCDMA with the Chinese actors (Seo and Koek, 2012), it was not surprised that they had a strategy to persuade the Chinese actors to join their value network.

Formation of Standard-Setting

Without a standard-setting organization, it was difficult to standardize a technology in the mobile communications industry. There was a well-established organization, 3GPP that had promoted technologies evolved from GSM such as GPRS, EDGE, and WCDMA. Therefore, it was natural for Ericsson and Nokia Siemens Networks to standardize LTE through 3GPP.

Openness of IPR (Intellectual Property Rights)

As equipment manufacturers, Ericsson and Nokia Siemens Networks did not base their business models on collecting royalty fees. Their interests were standardizing the LTE technology as a global 4G mobile communications technology standard. More LTE was adopted, more markets there were for them to sell their equipment. Thus, their IPR strategies were cross-licensing their patents with others to access other necessary patents and licensing their patents to others under FRAND (fair, reasonable and non-discriminatory) conditions.

6. EMERGENCE OF THE 4G MOBILE COMMUNICATIONS STANDARDS

Unfortunately, no mobile service provider had announced to adopt UMB. Most mobile service providers who had the CDMA2000 systems announced that they would adopt either WiMax or LTE for the next generation technology.

Qualcomm failed to persuade not only mobile service providers but also handset and system manufacturers. Those manufacturers were afraid of paying high license fees to Qualcomm by adopting the UMB technology because of Qualcomm's IPR strategy. This was not only reason why Qualcomm could not standardize the UMB technology, but it was true that Qualcomm had alienated many companies rather than made allies due to its IPR strategy. Without formulating a strong value network, Qualcomm finally announced that it would not pursue UMB any more in November 2008.

As WiMAX forum was first formed in June 2001, the WiMax camp started to develop and standardize the WiMax technology much earlier than others, considering the first commercial releases of CDMA2000 and WCDMA were in October 2000 and October 2001, respectively. While the first commercial WiMax service was provided over the networks of two Korean mobile communications service providers (SKT and KT) with the Samsung's involvement in June 2006, the first commercial LTE service was provided by TeliaSonera at Oslo and Stockholm in December 2009. More than three years gap opened an opportunity window to disseminate WiMax as the WiMax camp planned. Sprint-Nextel began to provide WiMax services in October 2008. Moreover, Clearwire that was largely owned by Sprint-Nextel, Comcast, Time Warner Cable, Bright House Networks, and Google adopted WiMax and offered mobile and fixed wireless broadband communications services. In addition, the fact that UMB was out of the competition arena was another opportunity for the WiMax camp to persuade mobile service providers whose networks were based on the CDMA2000 series.

During this period (between 2006 and 2009), the WiMax camp took the opportunity, so the future of WiMax looked rosy. However, observing the value network of WiMax, the participants were lopsided. It meant that there were not many actors from the mobile communications industry who participated in the WiMax value network except few actors such as Samsung and Sprint-Nextel, but more commitments from actors in the computer industry (e.g., Intel and Google) and the cable operating industry (e.g., Comcast, Time Warner Cable, and Bright House Networks). Considering Samsung as a handset manufacturer also supported the LTE development and the Korean government had a tendency to divide its bets on more than one potential standard (e.g., CDMA2000 and WCDMA), the lack of actors from the mobile communications industry made the WiMax value network vulnerable.

The LTE camp did not overlook this vulnerability of the WiMax value network. The LTE camp more focused on re-enforcing its existing value network from the WCDMA camp. At the same time, it tried to persuade mobile service providers who had the CDMA2000 or TD-SCDMA related networks such as Verizon and China Mobile. Especially, to persuade Chinese mobile service providers who were growing quickly with the large subscribers, the LTE camp also developed another version of the original LTE with the Chinese actors and others. Interestingly, they also included Qualcomm in developing this different version

of the original LTE. The original LTE is based on Frequency Division Duplexing method, while the other version, so-called TD-LTE is based on Time Division Duplexing. The TD-LTE technology provided an easier migration path to mobile service providers that had the TD-SCDMA (Time Division Synchronous CDMA) networks than the original LTE. With these strategies in configuring its value network, the LTE camp was successful in convincing key mobile service providers (e.g., Verizon) to commit to adopt LTE even before the first commercialization of LTE. This effort also broke the value network of WiMax. Samsung and the Korean mobile service providers switched to embrace the LTE technology. In addition, Clearwire, one of the largest WiMax service providers, also started to adopt LTE, while not expanding the already installed WiMax regions.

Figure 1 shows that the WiMax camp started much earlier than the LTE camp to standardize its technology. The WiMax camp was able to gain first mover advantage before 2009. However, the strategies of the LTE camp in configuring a value network overcame the first mover advantage of the WiMax camp.

Figure 1. The timeline of development of the 4G standards

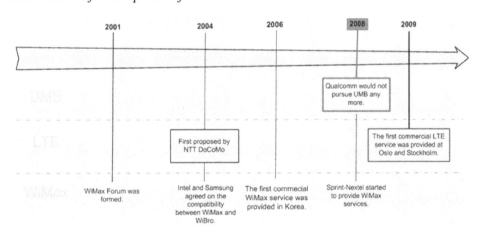

REFERENCES

Fitchard, K. (2010). Ericsson claims to hold a quarter of all LTE patents. *Connected Planet*. Retrieved September 1, 2012, from http://connectedplanetonline.com/3g4g/news/ericsson-lte-patents-061110/

Homeland Security News Wire. (2008). WiMAX patent alliance formed to promote the standard. *Homeland Security News Wire*. Retrieved September 1, 2012, from http://www.homelandsecuritynewswire.com/wimax-patent-alliance-formed-promote-standard

Seo, D., & Koek, J. W. (2012). Are Asian countries ready to lead a global ICT standardization? *International Journal of IT Standards and Standardization Research*, *10*(2), 29–44. doi:10.4018/jitsr.2012070103.

Chapter 9
Summary and Conclusion

EXECUTIVE SUMMARY

This last chapter provides the overall summary and conclusion of this book. This chapter explains why and how each chapter is included and inter-related. In addition, the limitations and contributions of this book are explained.

1. RECAPITULATION OF THEORY AND MODEL

The preface of this book introduced the overall theme and its significance. The general research question is: What is the nature of organizational strategies for technology standards? The concrete research question is how organizations reach and then adapt their strategies for standards before, during, and after the industry-wide standardization process.

The first chapter presented the background of standards strategy, starting with the general definition of standard and gave reasons for selecting the ICT (Information and Communications Technology) industry for this research. The meaning and importance of standards strategy were discussed, and three strands of literature on standards strategies in the ICT sector were reviewed. The first strand of literature is about the technological

configurations necessary to stimulate economic factors in technology standardization. The second strand of literature is about creating and maintaining the value networks required to develop the necessary technological configuration among actors. The literature herein identifies various types of collaborations among actors to create and maintain value networks. The third strand of literature is about the role of IPR (Intellectual Property Rights) – this is a significant issue for organizations, because organizations may use IPR as a tool to leverage their power in standardizing technologies. After the review, these three strands of literature were synthesized to draw out the three aspects of standards strategies – Configuration of Value Network, Formation of Standard-setting Organization, and Openness of IPR – that are fundamental to any investigation of organizational standards strategy.

As shown, technology standardizations are the result of agreements among actors including service providers, manufacturers, and govern-

DOI: 10.4018/978-1-4666-4074-0.ch009

ments. Technology standardization is not only about solving technical problems among systems, but also about strategic and political maneuvers and agreements among actors to gain competitive advantages. Thus, the context and configuration of technologies and markets largely constrain the actors' standards strategies, which in turn interactively shape the emerging technology standards and the market configuration they facilitate.

To analyze the formulation of organizational standards strategy and study how organizations' strategies interact and shape the evolution of technology standards, it is necessary to develop an appropriate conceptual model and theoretical framework. Chapter 2 began with reviewing and critiquing the existing theories and ended with introducing the more suitable Self-organized Complexity Unfolding Model based on combining the theories of ANT (Actor Network Theory) and Self-organized Complexity. Based upon this conceptual model, a Framework of Organizational Standards Strategy was developed to analyze individual organizations' standards strategy. This Framework of Organization Standards Strategy consists of three parts – (1) the organization's situation, (2) the organization's interpretation of its situation, and (3) the organization's formulation of its standards strategy. For part (1), based on the literature review, six elements were identified and proposed for modeling an organization's situation – (1) an organization's capabilities to meet market needs and opportunities, (2) the availability of complementary products or compatibility of products/services in the market, (3) the innovativeness of the technology involved, (4) the position of the organization in the market, (5) the availability of alternative or substitutable technologies in the market, and (6) the characteristics of intellectual property rights regarding the technology involved. For part (2), the perspective of "organization's striving to create and capture value" was used to trace, analyze, and understand how organizations interpret their situations. Lastly, for part (3), the three aspects introduced in Chapter 2 – namely,

Configuration of Value Network, Formation of Standard-setting Organization, and Openness of IPR – were used to analyze the formulation of organizational standards strategies given the organizational situations and interpretations.

Traditional theories of competitive advantage such as the Resource-based View and Competitive Forces are perhaps up to the task of analyzing standards strategies in simple and relatively stable industries such as the beverage and retail industries. However, high-tech industries such as mobile communications and personal computers are much more complex and dynamic. Firstly, there are many and diverse actors and many more factors to consider. Secondly, the factors influencing actors' situations, e.g. technology and product cycles, are continuously changing. Thus, a conceptual model and its associated theoretical framework for analyzing standards strategies in ICT need to be general and flexible enough to capture the changing characteristics of these complex and dynamic industries, yet specific enough to facilitate concrete description and analysis of the impact of organizational standards strategies on the development of the industries. This book illustrates that the Self-organized Complexity Unfolding Model and the Framework of Organizational Standards Strategy, as one approach to analyze these complex and dynamic industries, meet the criteria stated.

To my knowledge, this research is one of the first studies that tries to account for the complexity of real business environments at both the industrial level and the organizational level. Due to the exploratory nature of the research questions, it is important to validate the viability of the proposed model and framework in capturing the dynamic complexity of organizational standards strategies and industry structure. It was the challenge of doing truly large-scale case studies that covered more than 30 years of the mobile communications history. As to be seen, these large-scale longitudinal case studies clearly validate the viability of the proposed model and framework.

2. RECAPITULATION OF CASE STUDIES

To demonstrate how the proposed conceptual model and framework facilitate the descriptions and analyses of organizational standards strategies, Appendix introduced prototype case studies of IBM and NEC standards strategies in standardizing PC architecture. These cases were used because they are two of the most well known examples of standard-setting in information technology history. These prototype cases gave a first-cut illustration of how the proposed conceptual model and framework may be applied to analyze organizational standards strategies.

Applying the proposed conceptual model and framework to more complex cases was much more challenging than applying them to simple cases. Chapter 3 presented the reasons for selecting the evolution of technology standards in the mobile communications industry as the subject of the large-scale case studies. The complexity and importance of this industry were also discussed.

Chapter 4 started with a general background of the mobile communications industry including historical and technological backgrounds of each category of actors in the industry. This background information provided an overview of how the organizational context for formulating standards strategies was influenced by the existing industrial configuration.

After giving the background of the mobile communications industry, the proposed conceptual model and framework were applied to study the evolution of mobile technology standards in four chapters, tracing the unfolding of the standards from 1G, through 2G and 3G, to 4G.

Chapter 5 covered the emergence of 1G mobile communications technology standards. The standards strategies of three groups of major actors and markets were studies – the NMT group in Nordic Europe, AT&T in U.S.A, and the PTTs in France, West Germany and Italy. These cases illustrated how the different actors, when faced with vary-

ing situations, depending on their interpretations, could form different standards strategies. Chapter 5 also showed how, in spite of the fragmented communications markets, the interactions of these actors' strategies opened up opportunities for the NMT actors to become disproportionately important, relative to their size, in the emerging configuration of mobile communications.

The configuration of the value networks that gradually emerged out of the 1G mobile communications standardizations and development in the early 1980s, settled down and became the existing contextual and configuration, in the late 1980s, for the development of the 2G mobile communications technology standards. Chapter 6 began with the background of the 2G technology standards, including the technological innovation that enabled the migrations of market from 1G to 2G. This background is important for understanding the industrial configuration and organizational context in the development of the GSM and IS-95 standards. Chapter 6 analyzed the standards strategies of four major actors – the GSM group, West German and French governments, and Ericsson – who interacted to ultimately establish GSM as the pan-European 2G mobile communications technology standard. Chapter 6 also analyzed the standards strategies of three major actors – Qualcomm, PacTel, and the Korean government – who aligned and co-operated in establishing the IS-95 (CDMA) technology, as an alternative global 2G standard.

Although only one element – namely the market and political position – was different in the situations of the GSM group, of the West German and French governments, and of Ericsson, the standards strategies of these actors were very dissimilar, because they had very different interpretations of the situations and they had very different aspirations. The interactions of their divergent strategies drove the unfolding of the industrial alignment of the actors in standardizing GSM technology and the shaping of the value networks. The result, rather unexpected then but seeming quite inevitable after this analysis, was that the standard promoted by the relatively minor actor Ericsson won over the one promoted by the powerful French and German governments.

As for IS-95, the situations and strategies of the three main actors (Qualcomm, PacTel and Korean government) were quite different. However, in interpreting their situations they all perceived the opportunity to achieve their different goals through standardizing, commercializing, and diffusing the IS-95 technology. Thus, they were able to form a viable value network to standardize and develop the IS-95 technology. Because of this alignment, and because of the relative technological superiority of IS-95 over GSM, Qualcomm and the Korean manufacturers were able to carve out a small but not insignificant 2G market with the IS-95 standard, even though they were very late entrants to the global market. The successful commercialization and standardization of IS-95 as a 2G standard, which was very much in doubt then, turned out to significantly affect the emerging self-organized configuration of the future global mobile communications industry.

Chapter 5 and Chapter 6 demonstrated that the proposed conceptual model and framework were viable. Indeed they provided a logical and clear approach to investigate and explain very complex phenomena. The proposed model and framework showed how actors' situations were affected by the 1G industrial configuration; how these actors interpreted their situations and envisioned a future 2G mobile industry and market; how they developed standards strategies based on these interpretations and visions; and how the interactions of the value networks these actors tried to form based on their standards strategies influenced the self-organized configuration of the 2G mobile communications technology industry.

As actors headed towards 3G, they became more aware of the importance of technology standards, and became smarter in developing standards strategy to vie for control of the global market. This meant that actors tried to leverage their positions, to develop and exploit IPRs, and to maneuver to

obstruct opponents, etc., and they orchestrated their strategic moves in a more sophisticated fashion. Thus, the process of standardization of 3G mobile communications technology is much more complex than that of 2G. Chapter 7 applied the proposed conceptual model and framework to describe and explain the conflicts and reconciliations among five major actors (ETSI, NTT DoCoMo, Ericsson and Nokia, and Siemens) in their efforts to standardize WCDMA to protect the GSM market. Chapter 7 also studied the corresponding actions of actors from the CDMA camp (Qualcomm and Korean government) as they tried to standardize their technologies and reacted to the strategies of the GSM actors and NTT DoCoMo. In spite of the complexity of 3G standardization, the proposed conceptual model and framework were again able to successfully and clearly explain how actors developed their standards strategies and how the self-organized industrial configuration emerged from the interactions of the actors' strategies and actions.

Chapter 8 presented a very short discussion of the ongoing convergence between the telecommunications and computing industries. These discussions, though sketchy, aimed to point out that the struggle to set and benefit from the 4G mobile communications technology standard, which is just settled. Nevertheless, with judicious incorporation of the standards strategies of new actors from the computing industry, the model and framework proposed in this book served just as well for studying the unfolding of the 4G standard.

3. CONTRIBUTION OF THIS BOOK

The contributions of this book may be succinctly summarized as: (1) understanding how organizations determine their standards strategies; (2) suggesting the importance of organizational standards strategy as a way for organizations to gain and sustain competitive advantage; (3) highlighting the field of organizational standards strategy as an important part of overall business strategy; and (4) suggesting a holistic theoretical model and framework to analyze complex phenomena by integrating two existing theories (Self-Organized Complexity and Actor Network Theory).

As should be clear from the large scale longitudinal case studies from 1G through 4G in the evolution of mobile communications technology standards, it would have been impossible to explain clearly how the actors developed their standards strategies and how the industrial configuration evolved without a general, flexible, yet richly detailed conceptual model and framework similar to what is proposed in this research.

One of the limitations of the proposed framework is that it covers only certain basic elements or aspects of organizations' situations and interpretation of their situations. These elements or aspects do not cover all dimensions of organizational situation and interpretation. However, researchers may readily modify the proposed framework by refining and adding new factors relevant to their research questions.

Since each sector in ICT has its own unique characteristics, some of the situational elements in the proposed framework need to be redefined, depending on the case at hand. For example, the element of "availability of complementary products or compatibility of products/services" was quite clear and appropriate in the case of PC architecture standards, but it became a bit more vague in the case of the mobile communications industry, especially in the 1G and 2G periods. The problem in this case was how to define the core versus the complementary products. It turned out that for the mobile communications industry, the compatibility of products/services with existing and other systems was more appropriate for understanding organizational situation. Thus, generally speaking, to apply the proposed model and framework to ICT industries other than mobile communications industry, all the factors used for modeling "situations" and "interpretations" need to be re-evaluated for appropriateness.

This research is in its nature exploratory – it proposed an innovative model and framework appropriate for studying the standards strategies of organizations, which is an increasingly important part of the organizational strategy for survival and success in the complex and dynamic ICT industry. After validating the model and framework with large-scale case studies of the mobile communications industry, these model and framework will provide a sound basis for much further research. A short list of worthwhile future research would include:

1. Enrich and refine the case studies.
2. Apply the model and framework to study other ICT sectors.
3. After sufficiently extensive studies of various industries and situations, systematize the framework by doing a rigorous classification and organizational situations, interpretations, strategies, and the patterns of unfolding of standards and value networks.

Lastly, perhaps more importantly, the knowledge and understanding gained from the above future research could be synthesized to become a pragmatic prescriptive (normative) theory to aid organizations in their difficult task of formulating and pursuing successful standards strategies.

4. MANAGERIAL IMPLICATIONS

For managers, this book provides a way to assess and anlyze strategies of their own organizations as well as those of other organizations including competitors. In the industrial level, they can accordingly identify their organizatios' positions in standardizing a technology. There are books about mobile communications technologies and standards, but these books are descriptive, that describe the technological aspect of mobile communications technology, the general market development, or the discrete technology standardization. Two most important differences of this book are: first, it has an analytical model and framework based on Actor-Network Theory and Self-organized Complexity to analyze and explain organizations' strategies and behaviours in standardizing technology. This approach goes beyond the simple description of complex phenomena. This book provides a way to explain these complex phenomena clearly instead of describing the complexity as is. Another difference of this book is that it actually explains how the history of global mobile communications industry and market has unfolded by identifying major actors and analyzing their strategies and behaviours over 30 years from the first generation to the third generation. There is no book that explains this longitudinal aspect of mobile communications industry and market. Thus, this book can be useful for managers and strategists to evaluate their organizations' strategies and positions in standardizing a technology.

Appendix: Prototype Case Studies of IBM and NEC Standards Strategies

ABSTRACT

To demonstrate how the proposed framework facilitates the descriptions and analyses of organizational standards strategies, it is applied to one of the most well-known examples of standard setting in information technology history, the technology of PC (Personal Computer) architecture and the strategies of two different organizations involved in its standardization. The first organization is IBM, which introduced the first PC in 1981. The case study begins with an analysis of IBM's situation when it first released the PC, discusses IBM's interpretations of its situation based on the perspective of value creation and capture, then identifies three aspects of IBM's strategies – the configuration of value network, formation of standard-setting, and openness of IPR (Intellectual Property Rights). This case study ends with a discussion of what happened to IBM after the standardization of PC architecture. The second organization is NEC, which became the dominant player in the Japanese PC market and ruled the Japanese market throughout the 1980s and early 1990s. The change in NEC's fate after the introduction of Microsoft Windows 3.1 is also discussed. After analyzing these two cases, the two companies' strategies for standardization based on their different situations and interpretations are compared. Before moving to the cases, it is worth mentioning that various journals, books, and articles have been used to collect facts, and then applied to the framework proposed in this book. In particular, the book, *Standards Strategy and Policy* by Peter Grindley (1995), was very helpful for the IBM case and the article, "Innovation and Control in Standards Architectures: The Rise and Fall of Japan's PC-98" by West and Dedrick (2000) was useful for the NEC case.

1. THE PC ARCHITECTURE STANDARD IN THE USA

1.1. The Existing Configuration of the Industry Before Standardization

IBM was the dominant player in the global market for mainframe computers from the 1960s through the 1980s. Its market share in the U.S. mainframe market was once over 80%. Mainframe computers were expensive and highly profitable products mainly used by governments and large companies, which made IBM not very interested in the individual-level, so-called "microcomputer," which had very limited computing capacity.

The microcomputer market was launched in the 1970s with products introduced by small companies such as Micro Instrumentation and Telemetry Systems (MITS), Apple, and Commodore International. In the early stages, the microcomputer market was small and fragmented by these various companies with their computer architectures, so their products were not compatible with each other (Grindley, 1995). The value networks of these early microcomputers were vertically integrated, with each company developing in-house all the different components of its microcomputer. Thus, each company's microcomputer had its own OS (Operating Systems) and software applications, which meant that any single company lacked a sufficiently large installed base to lock-in complementary product manufacturers or customers and make its microcomputer into a standard.

The main actors at this stage were these vertical suppliers and customers who demanded microcomputers. We can describe the existing self-organized configuration thus: (1) actors like Micro Instrumentation and Telemetry Systems (MITS), Apple, and Commodore International formed vertically integrated networks that produced incompatible microcomputers. (2) They competed against each other to gain more market share, but all of them lacked a sufficiently large installed-base to stimulate economic factors like the bandwagon effect to standardize their technologies (Figure 1). (3) New PC customers were emerging, in contrast to the existing traditional mainframe customers, which implied new market needs for PC. This existing industrial configuration affected IBM's situation.

Although the microcomputer market was not as attractive as the mainframe computer market, IBM eventually perceived the possible opportunities in serving individual customers. In addition, the microcomputer technology was not an alternative or substitutable technology to the mainframe computer, making it an entirely different market for IBM. Therefore, IBM could bring its strong brand name into this new market without cannibalizing its mainframe market.

Once IBM decided to develop a microcomputer, it found that its experience and technology capabilities developed in the mainframe computer industry were helpful because the technology involved in developing a microcomputer was a continuous, not disruptive, innovation from that of the mainframe computer technology. However, in its rush to develop a microcomputer without draining resources from its more lucrative mainframe production, IBM decided that it would not vertically integrate production but instead chose to outsource the manufacturing of components, including the processing chip, operating system, and distributors, in order to expedite the time to introduce its microcomputer into the market (Grindley, 1995). IBM had the rights to manufacture Intel chips, because it used Intel 8086 for one of its products in exchange for allowing Intel the rights to use IBM's bubble memory technology, so IBM

Figure 1. The industrial configuration before PC architecture standardization

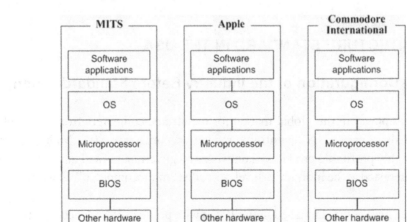

adopted Intel chips for its microcomputer. IBM also talked to Bill Gates for outsourcing its operating systems when Microsoft was a small, unknown software company, and used Sears and Computerland as distributors (Bellis, 2007).

This horizontal value network was different from the vertical value network that IBM had in place for its mainframe computer market, and different from vertical value networks of the other microcomputer companies. What IBM interpreted about its situation at the time was that the main value in the microcomputer market did not lie in microprocessor chips, OS, or distribution, but in fast penetration of the market with an affordable computer bearing the IBM brand name. IBM believed that it could still control the value network because of its role as system integrator and its proprietary BIOS technology, even though it adopted an open architecture for its microcomputer in the horizontal value network (West and Dedrick, 2000). This was the way to control networks and to influence the emerging self-organization of the industry during and after standardization. IBM finally released its first microcomputer, "the IBM PC," to the market in 1981. IBM's strategies of using an open PC architecture and creating the horizontal value network with the IBM brand name did indeed allow its PC to penetrate the market quickly.

1.2. Elements of IBM's Situation

We can analyze IBM's situation in deciding to launch a PC through the lenses of the six elements proposed in this research model.

- **Organization's Capabilities to Meet Market Needs and Opportunities:** There were market needs and opportunities in the market for microcomputers for individual and small business customers in the 1970s, but it was not so attractive to the mainframe manufacturing leader, IBM, at the time, while other companies like Apple developed and introduced their microcomputers to the market. Thus, IBM did not catch the potential value in the microcomputer market in its initial stages.

- **The Availability of Complementary Products or Compatibility of Products/Services:** During this time, there were vertically integrated microcomputers from different companies that were incompatible with each other. So the market remained small and segmented without locked-in complementary products. No one microcomputer gained enough of an installed base to create network externality or lock-in effects. This opened an opportunity for IBM to stimulate economic factors necessary for standardization, even though it was a later entrant.

- **The Type of Technology Innovation:** The microcomputer technology was not a disruptive innovation but a continuous one drawing from mainframe computer technology. Thus, IBM had enough capabilities to develop a microcomputer.

- **The Position of IBM in the Market:** IBM had a strong market position as a mainframe manufacturer, but it was under time pressure as a relatively later entrant to the microcomputer market.

- **The Availability of Alternative or Substitutable Technologies:** There were no alternative or substitute technologies for the microcomputer at the time. A mainframe computer was not an alternative or substitute to the microcomputer, because the mainframe computer had greater capabilities than the microcomputer and target customers were large companies rather than small businesses or individuals.

- **Ownership and Deployment of IP (Intellectual Property):** IBM owned the intellectual property rights for BIOS technology, which IBM considered an essential component of its PC. However, IBM neglected the importance of IPR control for Operating Systems.

1.3. IBM's Interpretations of its Situation

The microcomputer market was not initially attractive to IBM, that dominated the much more profitable mainframe computer market. IBM basically perceived little value in the microcomputer market compared to the mainframe computer market. Therefore, although IBM had enough resources and capabilities to develop and introduce its own microcomputer at any time, it did not pursue this strategy at the beginning stage of the microcomputer market.

IBM began to realize possible value in the microcomputer market as a new market independent from the mainframe computer market after the microcomputer market began to be developed by other companies. At this point, IBM perceived the value in producing a microcomputer with competitive pricing and many complementary products (peripherals and software) under a strong brand name, which IBM could do easily and rapidly by leveraging a small part of its resources and capabilities. Due to the fact that significant economic factors had not led to standardization of any one microcomputer technology when IBM decided to enter the market, it was still possible for IBM to dominate the market even though it entered the market much later than others.

IBM decided early on that the most effective route to standardization in the microcomputer market lay in having as many compatible products as possible (Grindley, 1995). In order to standardize IBM's PC in the market, complementary product manufacturers needed to adopt IBM PC architecture and develop peripherals and software products compatible to it. Therefore, IBM made the technical specifications of its machine widely available to encourage others to produce IBM PC compatible products as quickly as possible.

1.4. IBM's Strategies

Looking at IBM's situation and its interpretation of its situation, we can analyze its strategies from the three proposed aspects.

- **Configuration of the Value Network:** As a later entrant to the market working under time pressure, IBM created a horizontally configured value network to get its PC to the market as quickly as possible, outsourcing the OS from Microsoft, 16-bit microprocessor chips from Intel, and distribution through Sears and Computerland, which was different from its existing vertical network for mainframe computers. IBM also encouraged complementary product manufacturers by opening up the technical specifications for its PC in order to gain a large installed-base to spur standardization. The open architecture of IBM's PC, based on a 16-bit microprocessor, eventually edged out various mutually incompatible 8-bit microcomputers.
- **Formation of Standard-Setting:** As explained previously, companies at the time generally had vertically integrated value networks for production of their goods. In that period, formal standard-setting organizations for information technology products were rare in the U.S. Companies cooperated with others, but it was usually in the context of a buyer-seller relationship, different from today's strategic alliances. One large company would have vertical integration and small companies supplied what the large company required, like the relationship between IBM and Microsoft. So the usual strategy for information technology companies at the time was to make their products *de facto* standards by dominating the market. IBM's strategy for its PC was not exceptional.

- • **Openness of IPR (Intellectual Property Rights):** Although IBM adopted a horizontal value network and open PC architecture, the company still thought it could control the whole value network through its proprietary BIOS technology. Thus, its strategies for intellectual property rights were to tightly close and control the BIOS technology, while opening up its architecture to encourage others to adopt it.

The existing self-organized configuration in the industry before IBM introduced its PC consisted of vertically integrated networks that produced incompatible computers by different microcomputer manufacturers that each owned a small segment of the market. Those segments were still too small to instigate standardization of any microcomputer technology. It is clear that IBM created and followed strategies that were completely rational according to its situation and interpretations. Rushing to introduce its PC in the market, IBM chose strategies of the horizontal value network and open architecture, which indeed attracted other manufacturers to develop complementary products such as software. The more complementary products were available, the more customers chose the IBM PC. This in turn stimulated more companies to join the bandwagon of IBM PC.

While IBM standardized its PC quickly with a horizontal value network (Figure 2), it was still able to control the network by holding the IPR for BIOS. IBM's entrance in the microcomputer market therefore impacted the existing self-organized configuration and interactions of actors and networks, leading to the emergence of a new self-organized configuration that centered on IBM's horizontal value network and pushed other value networks to the edges.

1.5. The Impact of IBM Post-Standardization: The New Configuration

Since IBM introduced the PC, the market has overwhelmingly adopted this technology as de facto standard with its horizontal value network. The IBM PC open architecture held over 97% of the global PC market in 2005 (Reimer, 2005). However, this dominance of the PC market did not contribute to IBM's fortune, because IBM ultimately lost control of the value network for PC. Consequently, IBM eventually sold its PC unit to China-based Lenovo Group in 2005 (Lemon, 2005).

Figure 2. The newly emerged self-organized configuration during the standardization of IBM PC

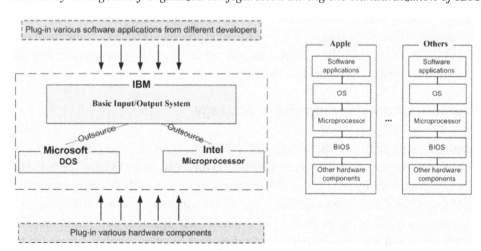

IBM became the dominant PC manufacturer as a result of the standardization of its PC, changing the self-organized configuration of the industry by forming a strong network with other actors. However, PC history did not unfold as IBM wished, even though it began so favorably, because the situation changed and the value network evolved in ways that IBM did not predict.

IBM's open architecture attracted not only many complementary product manufacturers, but also many IBM PC clone manufacturers. IBM did not realize that its proprietary BIOS could be copied through various reverse engineering methods (West and Dedrick, 2000). It tried to protect its rights, but court rulings allowed other companies to manufacture BIOS. Thus, BIOS became a commodity, or a product that is essentially undifferentiated and is interchangeable with another product of the same type (Chposky and Leonsis, 1988; Langlois, 1992).

Once BIOS was as widely available as MS-DOS and Intel microprocessors, PC assemblers could go out and buy all these major components from the market and manufacture PCs using IBM PC open architecture. There was no way for IBM to control its value network anymore, because IBM's value added role as system integrator was not unique. Indeed, many complementary product manufacturers, that witnessed the fast growth of the IBM PC, started to jump onto the bandwagon for standardization of the open architecture by supplying their products to the clone PC manufacturers. As hardware became stable and commoditized, the value of PCs shifted from hardware to software – and to having the stable OS and strong microprocessors which are necessary to run more complex software applications.

These circumstances created a great opportunity for Microsoft and Intel, that owned the IPR for these two significant PC components – OS and microprocessor respectively. Note that these two components were critical and non-substitutable, i.e. every PC clone must buy these components from Intel and Microsoft. They were open to use, but their designs were proprietary to Intel and Microsoft and closed to outside influence and control. This new situation allowed Intel and Microsoft to leverage their position as critical component suppliers to control the value network of the open PC architecture. The successful standardization of open PC architecture made both PC clone manufacturers as well as complementary product manufacturers locked-in to Microsoft OS and Intel microprocessors in a self-feeding cycle. At this point, the value network of IBM PC architecture was not a horizontal one controlled by IBM, rather it became a "*platform*" style controlled by Microsoft and Intel. Numerous personal computing machines could be launched on this WINTEL (Microsoft OS and Intel microprocessor) platform.

The strategies of these actors greatly impacted the self-organized configuration of the industry and IBM's situation in it. Now there was no significant difference between PCs produced by different PC makers using the WINTEL value network except price and distribution, which did not provide any competitive advantages to IBM. IBM started to lose its market share to other PC clone manufacturers that provided PCs at lower prices than IBM. IBM standardized its open IBM PC architecture successfully, but it failed to control it post-standardization (see Figure 3).

1.6. Conclusion: Understanding IBM's Strategy

The self-organized configuration of the industry evolved from vertically integrated networks of microcomputer manufacturers creating incompatible products in the early, small PC market, to the horizontally integrated network controlled by IBM that expanded the market drastically (Henderson and Clark 1990). However, IBM lost control of this emerged self-organized configuration when BIOS was developed by others through reverse-engineering methods. The emerged self-organized configuration opened a great situation for Microsoft and Intel to take over IBM's dominant position, evolving from the horizontally

Figure 3. The emerged self-organized configuration after PC architecture standardization

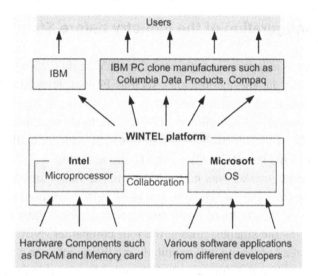

integrated network controlled by IBM to the WINTEL platform controlled by Microsoft and Intel, because the value of BIOS owned by IBM decreased while the value of OS and microprocessors increased.

IBM deliberately created strategies to make its PC dominant based on its situation and interpretation of the situation, but overestimated the value of BIOS and underestimated the value of OS and microprocessors. Consequently, it failed to control the value network. IBM's success in standardizing its PC in the market opened up space for other actors such as Microsoft, Intel and other IBM PC clone manufacturers to form their own networks that weakened IBM's network within the emerged self-organized configuration. These actors interpreted the situations and created strategies that benefited them, for example, other PC manufacturers could easily clone the IBM PC by buying components from Microsoft and Intel to assemble a PC. Due to the fact that Microsoft and Intel owned the IPR of OS and microprocessors respectively that were the essential components in a PC, they collaborated to develop and promote the WINTEL platform further. They also encouraged and supported these PC clone manufacturers to adopt their platform. Thus, the networks Microsoft and Intel formed with other actors made their positions stronger, while weakening IBM's network. The center of the self-organized configuration in the industry evolved one more time from IBM's vertical network to the WINTEL platform (see Figure 3).

IBM implemented its strategies intentionally in the way it wanted and was successful in standardizing its PC, but the self-organized configuration affected by its strategies changed over time and did not favor IBM post-standardization, because of the involvements of other actors and their interactions. As mentioned, each organization cannot form the self-organized configuration directly in the industry, but can only influence the evolution of the configuration through its interactions with others. To sustain an advantageous position and influence the evolution of the configuration more effectively, organizations need to build and maneuver their strategies not only during but also after standardization toward their desired value network configuration.

2. THE PC ARCHITECTURE STANDARD IN JAPAN

2.1. The Existing Configuration of the Industry before Standardization

The first computing machines in Japan were introduced by IBM in the 1950s (West and Dedrick, 2000). After this initial introduction, Japanese companies such as Fujitsu, Hitachi and NEC developed their own computers and penetrated the Japanese market quickly due to the Japanese government's policy of nurturing domestic companies (Anchordoguy, 1989). Due to these efforts, Fujitsu, Hitachi, NEC, and IBM Japan shared 85% of the Japanese mainframe market with mutually incompatible vertically integrated value networks in the late 1970s (West and Dedrick, 2000).

Similar to the U.S., the PC market was not attractive to the major Japanese mainframe computer manufacturers. Salespeople from Fujitsu, Hitachi, and IBM Japan even tried to redirect customers who were interested in low-priced PCs to more profitable minicomputers (Baba et al., 1996; Dedrick and Kraemer, 1998). However, as the smallest among the four companies, NEC perceived the PC market differently from others. Instead of seeing it as cannibalizing the mainframe computer market, NEC saw new possible market needs and opportunities in the PC market by observing the U.S. market and IBM's introduction of its PC architecture (West and Dedrick, 2000).

After NEC introduced its TK-80 as a do-it-yourself PC kit for hobbyists in 1976, Hitachi and Sharp released 8-bit PCs in 1978 (Basic Master MB-6880 and MZ-80K respectively), generally considered the first real PCs in Japan (West and Dedrick, 2000). NEC then introduced its own 8-bit PC, PC 8001. NEC quickly surpassed Hitachi and Sharp by penetrating the market using its existing distribution networks and held around 44% of the Japanese PC market from 1980 to 1982 (Kobayashi, 1986; West and Dedrick, 2000). However, the PC market as a whole was relatively lukewarm due to the limited capabilities of the 8-bit computers.

In these early stages of the Japanese microcomputer market, the main actors included not only the companies mentioned above, but also the Japanese government that set policies favoring domestic companies. The existing self-organized configuration for the Japanese microcomputer market consisted of vertically integrated networks by each manufacturer that produced incompatible products (Figure 4). The market did not have any dominant standardized PC, because the major actors aside from NEC had not perceived enough value in the PC market.

Pursuing these possible opportunities, NEC sought to adopt Microsoft's MS-DOS and asked the company to develop a Japanese-language version of its OS. Microsoft was too busy with the U.S. market to do so, but it allowed NEC to acquire the source code, modify it, and develop its own OS supporting Japanese characters (Boyd, 1997; West and Dedrick, 2000). This was how NEC tried to configure its value network with Microsoft at the beginning, but Microsoft's lack of availability gave NEC an opportunity to develop its own OS based on the Microsoft's source code. NEC also developed its own microprocessor chips based on the development resources and capabilities of its mainframe computer division and 8-bit PC (PC-8001) development (West and Dedrick, 2000). NEC then created its own BIOS to support Japanese character display (West and Dedrick, 2000). NEC's vertical integration of the value network for its new 16-bit PC allowed the company to make this new microcomputer to be backward compatible with existing 8-bit software applications.

Figure 4. The industrial configuration before PC standardization in Japan

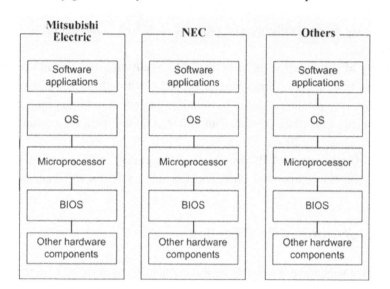

NEC finally released its first 16-bit PC (PC-9801) in Japan 14 months later than the introduction of the IBM PC in the U.S. but a year ahead of its competitors in the Japanese market. Although Mitsubishi Electric had introduced a 16-bit PC earlier, NEC, as one of the dominant players in the computing machinery market, became the real first mover in the Japanese PC market.

2.2. Elements of NEC's Situation

In reference to the existing PC market configuration in Japan, NEC's situation may be analyzed through the six proposed elements as follows:

- **Organization's Capabilities to Meet Market Needs and Opportunities:** Although the Japanese market was led by large companies and slower than the U.S. market, there were still significant market needs and opportunities in the market for PCs for individuals and small businesses in the late 1970s and the early 1980s. By observing the U.S. market, NEC understood the possible value in providing affordable computers, even with limited capabilities, for individual and small business customers.
- **The Availability of Complementary Products or Compatibility of Products/Services:** Although Mitsubishi Electric introduced a 16-bit PC in Japan first, when NEC introduced its version a year later there was still no real complementary product market for 16-bit PCs. By then there were U.S. manufacturers creating complementary products for the U.S. market, but the need to adapt products to handle the Japanese language was a barrier to entering the Japanese market. However, NEC had some complementary products for its new 16-bit PCs through backward compatibility with its dominant 8-bit PC.
- **The Type of Technology Innovation:** As mentioned in the IBM case, the PC technology was not a disruptive innovation. Therefore, companies that had developed a mainframe or minicomputer had the capabilities to develop and manufacture PCs.

- **The Position of NEC in the Market:** NEC's market position in Japan was not super-dominant like IBM's in the U.S. As the smallest among the four main players (Fujitsu, Hitachi, IBM Japan, and NEC) in the Japanese mainframe computer market, NEC's market position was strong but not the most superior.

- **The Availability of Alternative or Substitutable Technologies:** The mainframe computer was not an alternative or substitute technology to the PC. Although NEC's competitors perceived the PC as a supplementary product ("dumb terminal") to connect to large computers, NEC perceived it as more than a supplementary product, but rather as an independent product for an entirely new market (West and Dedrick, 2000).

- **Ownership and Deployment of IP (Intellectual Property):** NEC owned IPR for all three major components (BIOS, microprocessor chip, and OS) for its 16-bit PC, because it developed them in-house.

2.3. NEC's Interpretations of its Situation

Although the PC market was not as profitable as the mainframe market, NEC perceived potential value for 16-bit PCs by observing the development of the PC market in the U.S., while its competitors underestimated the possibilities of the PC market and tried to redirect their customers to more profitable minicomputers. Therefore, NEC could influence the formation of PC market by introducing a 16-bit PC earlier than its competitors. As a first mover, NEC took advantage of the opportunity to standardize its product by securing lots of complementary products. Thus, NEC's early interpretation of potential value in the PC market led to the timely introduction of a 16-bit PC with superior capabilities like Japanese language supportability and affordable price, which brought great value for customers.

Like IBM, NEC also determined that standardization of its 16-bit PC-9801 depended on gaining enough of an installed base to stimulate externalities and bandwagon effects. To gain a sufficiently large installed base, NEC needed complementary products. For these reasons, NEC first made sure that its 16-bit PC had backward compatibility with its 8-bit PC, so existing complementary products were available for the new product, even though they were not the kind of sophisticated software applications requiring 16-bit computing capability. NEC also distributed free computers to third-party software application developers with detailed specifications for the PC-9801 prior to release to encourage development of products compatible to the new computer. These activities derived from NEC's interpretation of the potential value in the PC market and the way to add value by supporting complementary product manufacturers.

2.4. NEC's Strategies

Reviewing NEC's situation and interpretation of its situation, we can analyze its strategies according to the three proposed aspects.

- **Configuration of the Value Network:** Although NEC tried to adopt Microsoft DOS for its 16-bit PC, the language difference gave NEC the opportunity to develop its own OS. IBM BIOS also needed to be modified to support the Japanese language, so NEC managed to develop its BIOS through reverse engineering without infringing on IBM's property rights. This situation triggered NEC to develop its own OS, BIOS, and microprocessor. NEC developed all the significant components in-house and integrated its value network for PC market vertically as it did for mainframe computers.

- **Formation of Standard Setting:** Japanese companies had a strong culture of vertically integrated value networks, even more so than U.S. computer companies. Some Japanese conglomerates today still control their entire supply chains and value networks, such as Toyota, Hitachi, Toshiba, Fujitsu, Matsushita, and Sony (Miyashita and Russell, 1994; Seo and Evaristo, 2006). Like the U.S., there was no tradition of a formal standard-setting process for information technology products. Rather, products became standards simply by dominating the market. Japan's corporate culture of vertical value networks and NEC's own existing value network for mainframe computer and PC markets contributed to the lack of a need for a formal standard-setting process. NEC simply tried to introduce its 16-bit PC earlier than its competitors in the Japanese market and standardize it through market competition, which is *de facto* standardization.
- **Openness of IPR:** As a first mover in the 16-bit PC market and the developer of its own OS, BIOS and microprocessor, NEC did not see the need to open its IPR for these technologies. Therefore, NEC kept all its intellectual property rights for the OS, BIOS, and microprocessor of its PC-9801.

With NEC's interpretation of the potential value that it could create for a PC market and consequent aggressive development, NEC was able to release its 16-bit PC, PC-9801, earlier than its competitors. NEC's early entrance into the market, backward compatibility of its 16-bit PC, and aggressive support for complementary product manufacturers allowed NEC to penetrate the market quickly, gaining 80% of the 16-bit PC market in the first year (West and Dedrick, 2000). Due to supporting complementary product manufacturers and fast market penetration, there were 3,589 software applications compatible with the PC-9801 by 1987, almost 10 times as many as those for its leading competitor, Fujitsu (Fransman, 1995; West and Dedrick, 2000). NEC's aggressive movement in the Japanese PC market affected the existing self-organized configuration, making the slowly developing Japanese PC market expand drastically. The NEC PC became the standard PC in the Japanese market as the IBM PC did in the U.S. NEC's strong network became the center of the newly emerged self-organized configuration, but all networks were still vertically integrated and produced products incompatible to each other (Figure 5).

2.5. The Impact of NEC Post-Standardization: The New Configuration

While IBM standardized PC architecture but eventually lost the market, NEC standardized its PC-9801 architecture successfully and was the leader in the Japanese PC market with 40-50% of the market until 1995. NEC was able to control its value network, while IBM failed to do so. Why was that the case? The major differences in their strategies were NEC's vertical value network and retention of the IPR for the major components for its PC, which together defined the value network. NEC was able to hold IPR by developing them in-house, which led to the vertical integration of its value network. As NEC's PC became more dominant, complementary product manufacturers became increasingly locked-in to the NEC PC. Other PC makers could not make clones of the NEC PC to take advantage of the great number of NEC PC compatible complementary products, because all significant PC components were owned by NEC. In addition, it was not attractive for complementary product manufacturers to manufacture products for PCs with much lower market share. This meant that once NEC built its network that won over others, more complementary product manufacturers produced more complementary products (Cottrell, 1996; West and Dedrick, 2000), making NEC's network even stronger in the emerged self-organized configuration.

NEC enjoyed its dominant position in the Japanese PC market for over a decade by standardizing its PC architecture while controlling the value network. Its heyday ended when Microsoft introduced Windows

190

Figure 5. The emerged self-organized configuration during and after PC standardization in Japan

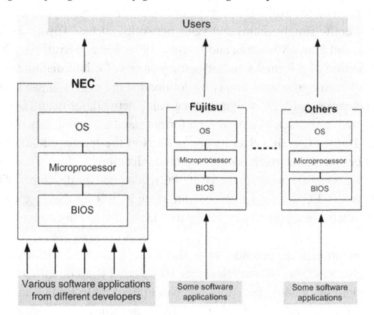

3.1, which could be run on all the different OS used at the time (West and Dedrick, 2000). Microsoft did this deliberately, seeking the same kind of OS-dominance in non-U.S. markets that it had achieved with the WINTEL platform in the U.S. Because of Windows' compatibility with different microprocessors, software application developers could develop one product for Windows 3.1 instead of different versions for the different OS used by each PC maker. Software application developers thus began to produce Windows-compatible products that could be used on all PCs regardless of the underlying architecture (see Figure 6). NEC's competitors saw the obvious value in this and began to adopt Windows 3.1. Because of this shift in software architecture, NEC's large installed base for complementary products became obsolete and its value network was not as effective. The distinction between the vertical value networks owned by different PC manufacturers did not matter anymore on the Windows platform, because any vertical value network could plug in to the Windows platform. The rate of collapse of NEC territory in the PC market increased as its competitors adopted Windows 3.1 more and more rapidly.

2.6. Conclusion: Understanding NEC's Strategy

The self-organized configuration in the Japanese PC market started similar to that of the U.S. PC market. The vertically integrated networks of each manufacturer producing incompatible PCs formed a small and segmented market. The PC market expanded greatly with NEC's vertically integrated value network at the center of the emerging self-organized configuration. NEC's early interpretation of the potential value in the PC market given its own situation made it possible to develop successful and sustainable strategies after as well as during standardization. Its strategies affected the evolution of the self-organized configuration, especially the strategy of holding the necessary IPR for the essential components in PC. This allowed NEC to capture the value of its product, while keeping competitors away from cloning its product.

Figure 6. The emerged self-organized configuration in Japan after Windows 3.1

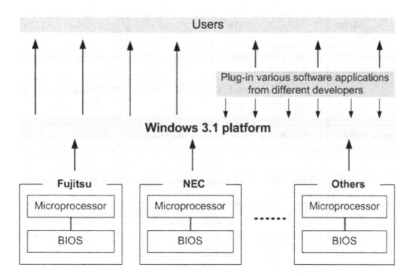

This configuration created a beneficial situation for NEC until the introduction of Windows 3.1, when the strategies of other actors changed the center of the self-organized configuration from NEC's network to the Windows 3.1 platform. As in the U.S., the new self-organized configuration of the Japanese PC market centered on the OS, not the PC hardware. So although NEC was able to hold on to its dominant position for much longer than IBM, it eventually met the same fate.

3. COMPARISON OF IBM'S AND NEC'S STRATEGIES

In comparing the cases of IBM in the U.S. and NEC in Japan, we can find both similarities and differences in their situations and their interpretations of their situation that influenced their strategies. Different strategies for standardization between IBM and NEC interestingly brought contrary consequences.

3.1. Comparison of Actors' Situations (IBM vs. NEC)

The most significant difference in the situations of IBM and NEC was the timing of their entrance into their respective markets. Another important difference was the IPR of technologies and the availability of complementary products or compatibility of products/services based on the language barrier. The language barrier gave NEC the opportunity to develop PC components in-house, because the language barrier blocked U.S. complementary product manufacturers from entering the market. Thus, NEC had the chance to develop 16-bit PC components and made complementary products for its 8-bit PC compatible to the PC-9801 (16-bit). As a result, IBM owned IPR on only BIOS, while NEC had the IPR of three essential components (BIOS, OS, and microprocessor).

192

Table 1. Comparison of actors' situations (IBM vs. NEC)

Actor / Element	IBM	NEC
Organization's capabilities to meet market needs and opportunities	(S) There were possible market needs and opportunities for PCs targeting individuals or small businesses.	
	(D) PC market in the U.S. was the first PC market globally.	(D) The Japanese PC market was slower than the U.S. market. NEC observed the U.S. market.
The availability of complementary products or compatibility of products/ services	(S) Existing PCs were vertically integrated and incompatible to each other. No one company had many more complementary products than others and the market was rather segmented.	
	(D) There were no existing complementary products, because the 16-bit IBM PC was the first PC product for IBM.	(D) NEC had some complementary products from backward compatibility with its 8-bit PC.
The type of technology innovation	(S) PC technology was not disruptive innovation compare to mainframe computer technology.	
Market position	(S) Both IBM and NEC were dominant players in their respective mainframe computer markets.	
	(D) IBM was super-dominant in the mainframe computer market, but had not entered the PC market.	(D) NEC was the smallest among the four dominant companies in the mainframe computer market, but quickly became dominant in the 8-bit PC market.
The availability of alternative or substitutable technologies	(S) There were no alternative or substitute technologies for the PC. The mainframe computer was not an alternative or substitute, because it had much greater capabilities and served different target customers.	
Ownership and deployment of IP (Intellectual Property)	(D) IBM owned the IPR of BIOS technology, but opened up the PC architecture.	(D) NEC owned IPR for all three major components: OS, BIOS and microprocessor.

[(S) Similarity, (D) Difference]

3.2. Comparison of Actors' Interpretations (IBM vs. NEC)

These different situations influenced their interpretations. One big difference between IBM and NEC was that NEC recognized market needs and opportunities in the Japanese market earlier than its competitors, while IBM recognized market needs and opportunities later than other microcomputer makers. The different market positions and regional locations influenced their interpretation of their different situations. This difference allowed NEC to understand the potential value it could create for the PC market earlier, while IBM underestimated the possible value in the PC market.

Thus the main difference in IBM's and NEC's interpretation, based on their different situations, was about the source of the potential value in the PC market. For IBM as a follower, value lay in speedy catch-up entrance to the market and branding: the company's priority was to introduce a competitive product under its name as quickly as possible, for which it was willing to sacrifice a tightly controlled vertical value network. NEC, on the other hand, saw the source of value in controlling the new PC market. As a first mover, it interpreted that value lay in introducing a PC earlier than others and securing more complementary products to penetrate the market earlier and faster.

3.3. Comparison of Actors' Strategies (IBM vs. NEC)

Based on the differences in situations and interpretations between IBM and NEC, they created different strategies for standardizing their products.

- **Configuration of the Value Network:** This was the critical difference between IBM and NEC strategies. IBM as a later entrant working under time pressure decided to outsource the OS, microprocessor, and distribution, creating a horizontal value network, while NEC as an early entrant taking advantage of the language barrier was able to develop its own major PC components and a vertically integrated value network for PC-9801 (see Figure 7).
- **Formation of Standard-Setting:** There was no difference in the formation of standard-setting for both IBM and NEC. At the time, organizations tried to standardize information technology products through market competition, creating *de facto* standards.
- **Openness of IPR:** IBM had IPR for BIOS, which it believed would allow it to control the whole value network, while NEC had IPR for BIOS, OS, and microprocessor through developing them in-house.

Although IBM's and NEC's strategies were different, both products penetrated their respective markets quickly, successfully creating a large installed base. This impacted the existing self-organized configurations, so their networks became the centers of newly emerged self-organized configurations by winning over other networks formed by their competitors. Even as a later entrant, IBM was able to standardize its PC rapidly by opening up its architecture and forming a horizontal value network with other actors, while NEC took advantage of its position as a first mover and through its tightly controlled value network and IPR.

While the differences in their strategies worked to standardize the products in the short term, these differences affected the fates of their products significantly in the next phase. IBM's strategies of having a horizontal value network and only holding the IPR for BIOS made its network vulnerable once BIOS became a commodity. Other actors, especially Microsoft and Intel, did not miss this weak point in IBM's network and interpreted this situation to develop strategies that could compete and push IBM's network aside. This made IBM's position of having the largest number of complementary products obsolete. The

Figure 7. IBM and NEC value networks

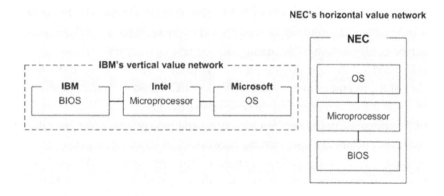

newly emerged self-organized configuration was not favorable to IBM anymore and left IBM's network on the periphery while putting the networks of Microsoft and Intel at the center. However, NEC sustained its position, because its strategies of the vertical value network and closely held ownership of IPR for the PC's essential components prevented other actors from developing clones of NEC's PC. Therefore, NEC sustained its position of having a dominant market share with the greatest numbers of complementary products that attracted more customers and complementary product manufacturers to join its network. NEC was able to hold on to its market significantly longer than IBM, but then also succumbed to unavoidable changes in the self-organized configuration due to the actions of other organizations, Microsoft in this case once again.

4. CONCLUSION OF PROTOTYPE CASE STUDIES

These prototype cases of IBM and NEC provide a first cut exploration for the usefulness of the proposed framework in this book. The first major theme of the framework is that organizational standards strategy is not arbitrarily formed but produced as a result of organizational situation and interpretation of the situation. IBM's situation as the incumbent in the mainframe computer market led the company to fail to recognize the potential value in the microcomputer. This made IBM come under time pressure when it finally understood the potential value, but it did rise to the occasion and bring a product to the market quickly through the strategy of developing a horizontal value network. IBM was still able to control the value network, because it held IPR for BIOS, which was the most valuable technology in a PC when the value of OS and microprocessor was low.

NEC's strategies again came out of its situation and interpretation of the situation. NEC was the least dominant among the four leading companies in the Japanese mainframe market, but it acknowledged the potential value in the microcomputer market early on by observing the U.S. PC market. NEC found itself in a situation where U.S. complementary product manufacturers were blocked out of the Japanese market by the language barrier. This fact, together with NEC's interpretation of the potential value in the microcomputer market affected its strategies, leading it to develop all essential components in-house to introduce the PC earlier than its competitors.

The second major theme of the framework is that there is a recursive (feed-forward and feed-back) relationship between the interactions of various organizations based on their strategies and the self-organized configuration of the industry. Organizations create their standards strategy based on the context of their situation in the existing self-organized configuration and their interpretation of the situation. Organizational implementation of strategy impacts the existing self-organized configuration, creating a new situation that all actors again interpret to create different strategies. The beginning situations of IBM and NEC as they set out to standardize their PCs in their respective markets were formed by the existing self-organized configuration of the mainframe computer industry and the early microcomputer market. For example, IBM's dominant position in the mainframe market made it overlook and underestimate the possibility of the microcomputer market, while NEC's least dominant position in the Japanese mainframe market made it look for other possibilities more actively.

IBM's and NEC's strategies, based on their situations and interpretations of the situations, provided ways to interact with other actors and form strong networks that shook the existing configuration of the industry and stimulated a new configuration to emerge. The networks of IBM and NEC were central in this newly emerged self-organized configuration. This emerged self-organized configuration again influenced the situations of actors. Actors interpreted their situation to develop strategies. The emerged self-organized configuration centered on IBM created the open architecture that IBM could not control. This left an opportunity for other actors such as Microsoft, Intel and IBM PC clone manufacturers, so those actors interpreted their situations and shaped their strategies to gain advantage. Actors' interactions based on their strategies again affected the self-organized configuration to evolve. Thus, this recursive relationship goes on.

Although the cases of NEC and IBM arose in two independent markets, the comparison of their cases reveals the significance of organizational strategies not only for practitioners but also for researchers. For academics, these cases provide a way to analyze organizational standards strategies in the evolution of dynamic and complex technology standardization, from a more holistic and pragmatic perspective that existing theories cannot provide. For example, according to Game Theory, IBM as the incumbent in the computing industry should not have had a horizontal value network, but it did. Economic theory does not explain why IBM would decline after it built enough installed base and kicked off the economic factors needed for standardization such as external network and bandwagon effect.

Although these two cases are relatively simple, with few main actors in two independent markets, it is difficult to understand the unfolding industrial configuration without Self-organized Unfolding Model. In particular, the framework unravels how complex organizational standards strategies arise and interact with others.

The following cases of the mobile communications technology standards are more complex, with more actors in interconnected markets tracked over a longer period of time. Therefore the hope is that the framework has been sufficiently explored and its explanatory power has been justified so that it can be usefully applied to this larger case study.

REFERENCES

Anchordoguy, M. (1989). *Computer inc*. Cambridge, MA: Harvard University Press.

Baba, Y., Takai, S., & Mizuta, Y. (1996). The user-driven evolution of the Japanese software industry: The case of customized software for mainframes. In Mowery, D. C. (Ed.), *The International Computer Software Industry: A Comparative Study of Industry Evolution and Structure*. New York, NY: Oxford University Press.

Bellis, M. (2007). *The history of the IBM PC*. Retrieved May 8, 2012, from http://inventors.about.com/library/weekly/aa031599.htm

Boyd, J. (1997). From chaos to competition: Japan's PC industry in transformation. *The Computer Journal*, 4(4), 29–32.

Chposky, J., & Leonsis, T. (1988). *Blue magic: The people, power and politics behind the IBM personal computer*. New York, NY: Facts on File, Inc..

Cottrell, T. (1996). Standards and the arrested development of Japan's microcomputer software industry. In Mowery, D. C. (Ed.), *The International Computer Software Industry: A Comparative Study of Industry Evolution and Structure*. New York, NY: Oxford University Press.

Dedrick, J., & Kraemer, K. L. (1998). *Asia's computer challenges: Threat or opportunity for the United States and the world?* New York: Oxford University Press.

Fransman, M. (1995). *Japan's computer and communications industry.* Oxford, UK: Oxford University Press.

Grindley, P. (1995). *Standards, strategy, and policy: Cases and stories.* Oxford: The University Press. doi:10.1093/acprof:oso/9780198288077.001.0001.

Henderson, R. M., & Clark, K. B. (1990). Architectural innovation: The reconfiguration of existing product rechnologies and the failure of established firms. *Administrative Science Quarterly, 35*(1), 9–30. doi:10.2307/2393549.

Kobayashi, K. (1986). *Computers and communications.* Cambridge, MA: MIT Press.

Langlois, R. (1992). External economies and economic progress: The case of the microcomputer industry. *Business History Review, 66*(1), 1–50. doi:10.2307/3117052.

Lemon, S. (2005). Lenovo completes purchase of IBM's PC unit. *IDG News Service.* Retrieved May 8, 2012, from http://www.pcworld.com/article/id,120670-page,1/article.html

Miyashita, K., & Russell, D. W. (1994). *Keiretsu: Inside the hidden Japanese conglomerates.* New York: McGraw-Hill, Inc..

Reimer, J. (2005). *Total share: 30 years of Personal computer market share figures.* Ars Technica, LLC.

Seo, D., & Evaristo, R. (2006, August). *The influence of national culture in shaping organizational forms in Korea, Italy, and Japan.* Paper presented at the Annual Meeting of Academy of Management. Atlanta, GA.

West, J., & Dedrick, J. (2000). Innovation and control in standards architectures: The rise and fall of Japan's PC-98. *Information Systems Research, 11*(2), 197–216. doi:10.1287/isre.11.2.197.11778.

Compilation of References

Adamson, W. (2004). How NTT DoCoMo stumbled in the race to 3G and its amazing recovery. *imodestrategy. com*. Retrieved April 10, 2008 from http://www.imodestrategy.com/2004/06/how_ntt_docomo_.html

Adamson, W. (2005). DoCoMo taps LG for roaming revenue expansion. *imodestrategy.com*. Retrieved April 10, 2008 from http://www.imodestrategy.com/2005/06/docomo_taps_lg_.html

Amabile, T. M. (1996). *Creativity in context: Update to the social psychology of creativity*. Boulder, CO: Westview Press.

Anderson, P. (1999). Complexity theory and organization science. *Organization Science*, *10*(3), 216–232. doi:10.1287/orsc.10.3.216.

Argote, L. (1999). *Organizational learning: Creating, retaining, and transferring knowledge*. Boston, MA: Kluwer Academic.

Argyris, C., & Schon, D. A. (1996). *Organizational learning II*. Boston, MA: Addison Wesley.

ARIB IMT-2000 Study Committee. (1998). *Japan's revised proposal for candidate radio transmission technology on IMT-2000: W-CDMA*. Japan: ARIB, Association of Radio Industries and Businesses.

Arthur, W. B. (1996). Increasing returns and the new world of business. *Harvard Business Review*, *74*(4), 100–109. PMID:10158472.

Ashby, W. R. (1947). Principles of the self-organizing dynamic system. *The Journal of General Psychology*, *37*, 125–128. doi:10.1080/00221309.1947.9918144 PMID:20270223.

Barney, J. B. (1991). Firm resources and sustained competitive advantage. *Journal of Management*, *17*(1), 99–120. doi:10.1177/014920639101700108.

Bekkers, R. (2001). *Mobile telecommunications standards: GSM, UMTS, TETRA, and ERMES*. Boston, MA: Artech House.

Benbya, H., & McKelvey, B. (2006). Using coevolutionary and complexity theories to improve IS alignment: A multi-level approach. *Journal of Information Technology*, *21*(4), 284–298. doi:10.1057/palgrave.jit.2000080.

Besen, S. M., & Farrell, J. (1994). Choosing how to compete: Strategies and tactics in standardization. *The Journal of Economic Perspectives*, *8*(2), 117–131. doi:10.1257/jep.8.2.117.

Blackman, C. R. (1998). Convergence between telecommunications and other media: How should regulation adapt? *Telecommunications Policy*, *22*(3), 163–170. doi:10.1016/S0308-5961(98)00003-2.

Blind, K., & Thumm, N. (2004). Interrelation between patenting and standardisation strategies: Empirical evidence and policy implications. *Research Policy*, *33*(10), 1583–1598. doi:10.1016/j.respol.2004.08.007.

Boar, B. H. (1984). *Application prototyping: A requirements definition strategy for the '80s*. New York, NY: John Wiley & Sons, Inc..

Bores, C., Saurina, C., & Torres, R. (2003). Technological convergence: A strategic perspective. *Technovation*, *23*(1), 1–13. doi:10.1016/S0166-4972(01)00094-3.

Bowman, C., & Ambrosini, V. (2000). Value creation versus value capture: Towards a coherent definition of value in strategy. *British Journal of Management*, *11*(1), 1–15. doi:10.1111/1467-8551.00147.

Boyd, J. (2008). *Observe-orient-decide-act*. Retrieved April 10, 2008 from http://www.d-n-i.net/dni/john-r-boyd/

Business Wire. (2002). QUALCOMM makes commitment for $200 million strategic investment in reliance communications limited, reliance plans to deploy nationwide CDMA network in India. *Business Wire*. Retrieved May 8, 2012, from http://findarticles.com/p/articles/mi_m0EIN/is_2002_Jan_10/ai_81561801

Calhoun, G. (1988). *Digital cellular radio*. Norwood, MA: Artech House.

Callon, M. (1986). Some elements of a sociology of translation: Domestication of the scallops and the fishermen. In Law, J. (Ed.), *Power, Action and Belief: A New Sociology of Knowledge?* London: Routledge & Kegan Paul.

Camazine, S., Deneubourg, J., Franks, N. R., Sneyd, J., Theraulaz, G., & Bonabeau, E. (Eds.). (2003). *Self-organization in biological systems*. Princeton, NJ: Princeton University Press.

Carlsson, B., & Stankiewicz, R. (1991). On the nature, function and composition of technological systems. *Journal of Evolutionary Economics*, *1*(2), 93–118. doi:10.1007/BF01224915.

Cattaneo, G. (1994). The making a pan-European network as a path-dependency process: The case of GSM versus IBM (integrated broadband communications) network. In Pogorel, G. (Ed.), *Global Telecommunications Strategies*. Amsterdam: Elsevier Science.

CDG. (2008). 3G CDMA worldwide diffusions. *CDMA Worldwide Database*. Retrieved April 10, 2008 from http://www.cdg.org/worldwide

Chandler, A. D. (1962). *Strategy and structure: Chapters in the history of the industrial enterprise*. Cambridge, MA: The MIT Press.

Chandler, A. D. (1990). *Scale and scope: The dynamics of industrial capitalism*. Cambridge, MA: Harvard University Press. doi:10.2307/3115503.

Channing, I. (1998). Crunch time for UMTS. *GSM World Focus*, 51-55

Chiesa, V., & Toletti, G. (2003). Standard-setting strategies in the multimedia sector. *International Journal of Innovation Management*, *7*(3), 281–308. doi:10.1142/S1363919603000829.

Christensen, C. M. (1997). *The innovator's dilemma: When new technologies cause great firms to fail*. Boston, MA: Harvard Business School Press.

Christensen, C. M., & Rosenbloom, R. S. (1995). Explaining the attacker's advantage: Technological paradigms, organizational dynamics, and the value network. *Research Policy*, *24*(2), 233–257. doi:10.1016/0048-7333(93)00764-K.

Clark, D. (2007). China misdials on mobiles. *Far Eastern Economic Review*, *170*(10), 52-57.

Cohen, W. M., Goto, A., Nagata, A., Nelson, R. R., & Walsh, J. P. (2002). R&D spillovers, patents and the incentives to innovate in Japan and the United States. *Research Policy*, *31*(8-9), 1349–1367. doi:10.1016/S0048-7333(02)00068-9.

Communications Week International. (1997, June 30). Euro backing boost for Japan's W-CDMA. *Communications Week International*, 25.

Cunha, E, M. P., & Da Cunha, J. V. (2006). Towards a complexity theory of strategy. *Management Decision*, *44*(7), 839–850. doi:10.1108/00251740610680550.

David, P. A. (1985). Clio and the economics of QWERTY. *The American Economic Review*, *75*(2), 332–337.

David, P. A. (1997). *Path dependence and the quest for historical economics: One more chorus of the ballad of QWERTY*. Oxford, UK: University of Oxford.

David, P. A., & Greenstein, S. (1990). The economics of compatibility standards: An introduction to recent research. *Economics of Innovation and New Technology*, *1*, 3–41. doi:10.1080/10438599000000002.

David, P. A., & Steinmueller, W. E. (1994). Economics of compatibility standards and competition in telecommunication networks. *Information Economics and Policy*, *6*(3-4), 217–241. doi:10.1016/0167-6245(94)90003-5.

Drazin, R., & Sandelands, L. (1992). Autogenesis: A perspective on the process of organizing. *Organization Science*, *3*(2), 230–249. doi:10.1287/orsc.3.2.230.

Estep, M. (2006). *Self-organizing natural intelligence: Issues of knowing, meaning, and complexity*. Dordrecht, The Netherlands: Springer.

ETSI. (1988). *Universal mobile telecommunications system (UMTS), concept groups for the definition of the UMTS terrestrial radio access (UTRA) (Report TR 101 397 v3.0.1)*. Sofia Antipolis, France: ETSI.

ETSI. (1996, July). *ETSI technical report 314*. Sofia Antipolis, France: ETSI.

ETSI. (1999). *ETSI SMG moves to the selection of a 3rd generation radio access system*. Sofia Antipolis, France: ETSI.

ETSI. (2008). IPR online database. *ETSI*. Retrieved May 8, 2012, from http://Webapp.etsi.org/ipr/

ETSI. (2008). *IPR online database*. Retrieved May 8, 2012, from http://Webapp.etsi.org/ipr/

Farley, T. (2008). *Telephone history series*. Retrieved May 8, 2012, from http://www.privateline.com/TelephoneHistory/History1.htm

Farrell, J., & Saloner, G. (1986). Installed base and compatibility: Innovation, product preannouncements, and predation. *The American Economic Review*, *76*(5), 940–955.

Faulkner, D., & Campbell, A. (2003). Introduction to volume 1: Competitive strategy through different lenses. In D. Faulkner & A. Campbell (Eds.), The Oxford Handbook of Strategy - Volume 1: A Strategy Overview and Competitive Strategy (pp. 1-17). New York: Oxford University Press.

Ferguson, R. B. (2006). RFID standard battle rages: HF or UHF? *eWEEK.com*. Retrieved from http://www.eweek.com

Ferguson, R. B. (2006). Wal-Mart's new CIO Says he'll back RFID. *eWEEK.com*. Retrieved from http://www.eweek.com

Fifield, P. (1992). *Marketing strategy*. Oxford, UK: Butterworth-Heinemann Ltd..

Fine, C. H. (1998). *Clockspeed: Winning industry control in the age of temporary advantage*. Reading, MA: Perseus Books.

Fitchard, K. (2010). Ericsson claims to hold a quarter of all LTE patents. *Connected Planet*. Retrieved September 1, 2012, from http://connectedplanetonline.com/3g4g/news/ericsson-lte-patents-061110/

Forum, U. M. T. S. (2008). UMTS commercial deployments. *Wireless Intelligence*. Retrieved May 8, 2012 from http://www.umts-forum.org/content/view/2000/98/

Frohlich, M. T., & Westbrook, R. (2002). Demand chain management in manufacturing and services: Web-based integration, drivers and performance. *Journal of Operations Management*, *20*(6), 729–745. doi:10.1016/S0272-6963(02)00037-2.

Gallagher, M. D., & Snyder, R. A. (1997). *Mobile telecommunications networking with IS-41*. New York: McGraw-Hill.

Gambardella, A., & Torrisi, S. (1998). Does technological convergence imply convergence in markets? Evidence from the electronics industry. *Research Policy*, *27*(5), 445–463. doi:10.1016/S0048-7333(98)00062-6.

Gao, P. (2005). Using actor-network theory to analyse strategy formulation. *Information Systems Journal*, *15*(3), 255–275. doi:10.1111/j.1365-2575.2005.00197.x.

Garrard, G. A. (1998). *Cellular communications: Worldwide market development*. Boston, MA: Artech House.

Goldbaum, D. (2006). Self-organization and the persistence of noise in financial markets. *Journal of Economic Dynamics & Control*, *30*(9/10), 1837–1855. doi:10.1016/j.jedc.2005.08.015.

Goold, M., & Luchs, K. (2003). Why diversify? Four decades of management thinking. In D. Faulkner & A. Campbell (Eds.), The Oxford Handbook of Strategy - Volume 2: Corporate Strategy (pp. 17-42). New York: Oxford University Press.

Graham, I., Spinardi, G., Williams, R., & Webster, J. (1995). The dynamics of EDI standards development. *Technology Analysis and Strategic Management*, *7*(1), 3–20. doi:10.1080/09537329508524192.

Grant, R. M. (1991). The resource-based theory of competitive advantage: Implications for strategy formulation. *California Management Review*, *33*(1), 114–135. doi:10.2307/41166664.

Grindley, P. (1995). *Standards, strategy, and policy: Cases and stories*. Oxford, UK: The University Press. doi:10.1093/acprof:oso/9780198288077.001.0001.

GSM MoU Association. (1999). *GSM MoU association chairman calls on manufacturers to end technology battle.* GSM MoU Association.

Gupta, P. (2008). *EDGE! Will TDMA and GSM ever meet? Mobile wireless communications tomorrow.* Retrieved May 8, 2012, from http://www.wirelessdevnet.com/channels/wireless/training/mobilewirelesstomorrow5.html

Haken, H. (2006). *Information and self-organization: A macroscopic approach to complex systems* (3rd ed.). New York: Springer.

Hanseth, O., Jacucci, E., Grisot, M., & Aanestad, M. (2006). Reflexive standardization: Side effects and complexity in standard making. *Management Information Systems Quarterly, 30,* 563–581.

Hilton, G. W. (1990). *American narrow gauge railroads.* Stanford, CA: Stanford University Press.

Holcombe, A. N. (1911). *Public ownership of telephones on the continent of Europe.* New York: Houghton Mifflin Co..

Holma, H., & Toskala, A. (2000). *WCDMA for UMTS: Radio access for third generation mobile communications.* New York: Wiley.

Homeland Security News Wire. (2008). WiMAX patent alliance formed to promote the standard. *Homeland Security News Wire.* Retrieved September 1, 2012, from http://www.homelandsecuritynewswire.com/wimax-patent-alliance-formed-promote-standard

India Telecom. (2003, January 1). LG gets $104 million order from reliance. *India Telecom.*

Iversen, E. (1999). Standardisation and intellectual property rights: Conflicts between innovation and diffusion in new telecommunications systems. In Jakobs, K. (Ed.), *Information Technology Standards and Standardization: A Global Perspective* (pp. 80–101). Hershey, PA: Idea Group. doi:10.4018/978-1-878289-70-4.ch006.

Kaghan, W. N., & Bowker, G. C. (2001). Out of machine age? Complexity, sociotechnical systems and actor network theory. *Journal of Engineering and Technology Management, 18*(3/4), 253–269. doi:10.1016/S0923-4748(01)00037-6.

Kalavakunta, R., & Kripalani, A. (2005, January). Evolution of mobile broadband access technologies and services - Considerations and solutions for smooth migration from 2G to 3G networks. In *Proceedings of the IEEE International Conference on Personal Wireless Communications (ICPWC),* (pp. 144-149). IEEE.

Kang, S. C., Morris, S. S., & Snell, S. A. (2007). Relational archetypes, organizational learning, and value creation: Extending the human resource architecture. *Academy of Management Review, 32*(1), 236–256. doi:10.5465/AMR.2007.23464060.

Katz, M. L., & Shapiro, C. (1985). Network externalities, competition, and compatibility. *The American Economic Review, 75*(3), 424–440.

Katz, M. L., & Shapiro, C. (1994). Systems competition and network effects. *The Journal of Economic Perspectives, 8*(2), 93–115. doi:10.1257/jep.8.2.93.

Kay, J., McKiernan, P., & Faulkner, D. (2003). The history of strategy and some thoughts about the future. In D. Faulkner & A. Campbell (Eds.), The Oxford Handbook of Strategy - Volume I: A Strategy Overview and Competitive Strategy (pp. 21-26). New York: Oxford University Press.

KDDI. (2008). *KDDI history.* Retrieved May 8, 2012, from http://www.kddi.com/english/corporate/kddi/history/index.html

King, J. L., & West, J. (2002). Ma Bell's orphan: US cellular telephony, 1947-1996. *Telecommunications Policy, 26*(3-4), 189–203. doi:10.1016/S0308-5961(02)00008-3.

Krugman, P. R. (1996). *The self-organizing economy.* Cambridge, MA: Blackwell Publishers.

LaForge, P. (2006). Race for the future: The 3G mobile migration. *Telephony, 247*(14), 2–8.

Latour, B. (1987). *Science in action: How to follow scientist and engineers through society.* Cambridge, MA: Harvard University Press.

Latour, B. (1996). *Aramis, or the love of technology.* Cambridge, MA: Harvard University Press.

Lea, G., & Hall, P. (2004). Standards and intellectual property rights: An economic and legal perspective. *Information Economics and Policy, 16*(1), 67–89. doi:10.1016/j.infoecopol.2003.09.005.

Lea, M., O'Shea, T., & Fung, P. (1995). Constructing the networked organization: Content and context in the development of electronic communications. *Organization Science, 6*(4), 462–478. doi:10.1287/orsc.6.4.462.

Lee, J. Y., & Chan, K. C. (2003). Assessing the operations innovation bandwagon effect: A market perspective on the returns. *Journal of Managerial Issues, 15*(1), 97–105.

Lee, W. C. Y. (2001). *Lee's essentials of wireless communications*. New York: McGraw-Hill.

Lehenkari, J., & Miettinen, R. (2002). Standardisation in the construction of a large technological system - The case of the Nordic mobile telephone system. *Telecommunications Policy, 26*(3-4), 109–127. doi:10.1016/S0308-5961(02)00004-6.

Lemley, M. A. (2002). Intellectual property rights and standard-setting organizations. *California Law Review, 90*(6), 1889–1980. doi:10.2307/3481437.

Lepak, D. P., Smith, K. G., & Taylor, M. S. (2007). Value creation and value capture: A multilevel perspective. *Academy of Management Review, 32*(1), 180–194. doi:10.5465/AMR.2007.23464011.

Liebowitz, S. J., & Margolis, S. E. (1990). The fable of the keys. *The Journal of Law & Economics, 33*(1), 1–25. doi:10.1086/467198.

Liebowitz, S. J., & Margolis, S. E. (1994). Network externality: An uncommon tragedy. *The Journal of Economic Perspectives, 8*(2), 133–150. doi:10.1257/jep.8.2.133.

Li, F., & Whalley, J. (2002). Deconstruction of the telecommunications industry: From value chains to value networks. *Telecommunications Policy, 26*(9-10), 451–472. doi:10.1016/S0308-5961(02)00056-3.

Lu, W. W. (2000). *China 3G: TD-SCDMA behind the Great Wall*. Paper presented at 3G'2000. San Francisco, CA.

Mak, K. T. (2006). *Standard strategy for high-tech business*. Chicago, IL: University of Illinois at Chicago.

Malik, O. (2007). India finally has a 3G plan: WiMax in the mix, too. *GIGaom*. Retrieved May 8, 2012, from http://gigaom.com/2007/11/13/india-finally-has-a-3g-plan-wimax-in-the-mix-too/

Manish, C. (2004). Minimize your expenses using reliance phones. *Reliance India Mobile Review*. Retrieved May 8, 2012, from http://www.mouthshut.com/review/Reliance_India_Mobile-55473-1.html

Manninen, A. T. (2002). *Elaboration of NMT and GSM standards: From idea to market*. (Academic Dissertation). University of Jyvaskyla, Jyvaskyla, Finland.

Markus, M. L., Steinfield, C. W., Wigand, R. T., & Minton, G. (2006). Industry-wide information systems standardization as collective action: The case of the U.S. residential mortgage industry. *Management Information Systems Quarterly, 30*, 439–465.

McGee, J., & Sammut Bonnici, T. A. (2002). Network industries in the new economy. *European Business Journal, 14*(3), 116–132.

McLaughlin, C. P., Pannesi, R. T., & Kathuria, N. (1991). The different operations strategy planning process for service operations. *International Journal of Operations & Production Management, 11*(3), 63–76. doi:10.1108/EUM0000000001268.

Merriam-Webster. (1998). *Webster's ninth new collegiate dictionary*. Springfield, MA: Merriam-Webster Inc..

Meurling, J., & Jeans, R. (1994). *The mobile phone book: The invention of the mobile phone industry*. London: Communications Week International on behalf of Ericsson Radio Systems.

Miller, A., & Dess, G. G. (1996). *Strategic management*. New York: McGraw-Hill.

Mobile Communications International. (1996, November). Japan shakes up for 3rd generation. *Mobile Communications International*, 16.

Mobile Europe. (1997, July/August). European commission calls for swift response on UMTS. *Mobile Europe*, 15.

Mock, D. (2005). The qualcomm equation: How a fledgling telecom company forged a new path to big profits and market dominance. New York: AMACOM (American Management Association).

Morel, B., & Ramanujam, R. (1999). Through the looking glass of complexity: The dynamics of organizations as adaptive and evolving systems. *Organization Science, 10*(3), 278–293. doi:10.1287/orsc.10.3.278.

Morse, J. (2007, June 15). China steps up 3G tempo. *WirelessWeek.*

Nandhini, V. (2008). *Migration of GSM networks to GPRS: Wipro technologies.* Retrieved May 8, 2012, from http://www.tslab.ssvl.kth.se/csd/projects/0308/gsmtogprs.pdf

Nikkei Electronics Asia. (2004, November). Qualcomm, NTT DoCoMo to promote global WCDMA deployment. *Nikkei Electronics Asia.* Retrieved May 8, 2012, from http://techon.nikkeibp.co.jp/NEA/archive/200411/343877/

Noam, E. M. (1992). *Telecommunications in Europe.* New York: Oxford University Press.

OECD. (2002). *Measuring the information economy.* Paris: OECD.

OECD. (2006). *Information technology outlook.* Paris: OECD.

OECD. (2006). *Science and technology.* Paris: OECD.

Oshri, I., & Weeber, C. (2006). Cooperation and competition standards-setting activities in the digitization era: The case of wireless information devices. *Technology Analysis and Strategic Management, 18*(2), 265–283. doi:10.1080/09537320600624196.

Palmberg, C. (2002). Technological systems and competent procurers - The transformation of Nokia and the Finnish telecom industry revisited? *Telecommunications Policy, 26*(3-4), 129–148. doi:10.1016/S0308-5961(02)00005-8.

Penrose, E. T. (1959). *The theory of the growth of the firm.* New York: Wiley.

Pilato, F. (2004, November 18). GSM takes 74% share of global wireless market. *Mobile Magazine.* Retrieved April 10, 2008 from http://www.mobilemag.com/content/100/104/C3425/

Porter, M. E. (1980). *Competitive strategy.* New York: The Free Press.

Porter, M. E. (1985). *Competitive advantage: Creating and sustaining superior performance.* New York: The Free Press.

Post, J. E., Preston, L. E., & Sachs, S. (2002). *Redefining the corporation: Stakeholder management and organizational wealth. Stanford, CA.* Stanford: Business Books.

Powell, W. W., & Snellman, K. (2004). The knowledge economy. *Annual Review of Sociology, 30,* 199–220. doi:10.1146/annurev.soc.29.010202.100037.

Priem, R. L. (2007). A consumer perspective on value creation. *Academy of Management Review, 32*(1), 219–235. doi:10.5465/AMR.2007.23464055.

Puffert, D. (2000). The standardization of track gauge on North American railways, 1830-1890. *The Journal of Economic History, 60*(4), 933–960.

Puffert, D. (2002). Path dependence in spatial networks: The standardization of railway track gauge. *Explorations in Economic History, 39*(3), 282–314. doi:10.1006/exeh.2002.0786.

Rivette, K. G., & Kline, D. (2000). Discovering new value in intellectual property. *Harvard Business Review, 78*(1), 54–66.

Rohracher, H. (2003). The role of users in the social shaping of environmental technologies, innovation. *European Journal of Soil Science, 16*(2), 177–192.

Rosen, B. N. (1994). The standard setter's dilemma: Standards and strategies for new technology in a dynamic environment. *Industrial Marketing Management, 23*(3), 181–190. doi:10.1016/0019-8501(94)90031-0.

Rudberg, M., & Olhager, J. (2003). Manufacturing networks and supply chains: An operations strategy perspective. *Omega, 31*(1), 29–39. doi:10.1016/S0305-0483(02)00063-4.

Sammut-Bonnici, T., & McGee, J. (2002). Network strategies for the new economy. *European Business Journal, 14*(4), 174–185.

Sarker, S., Sarker, S., & Sidorova, A. (2006). Understanding business process change failure: An actor-network perspective. *Journal of Management Information Systems, 23*(1), 51–86. doi:10.2753/MIS0742-1222230102.

Scherer, F. M., & Ross, D. (1990). *Industrial market structure and economic performance.* Boston, MA: Houghton Mifflin.

Schilling, M. A. (2002). Technology success and failure in winner-take-all markets: The impact of learning orientation, timing, and network externalities. *Academy of Management Journal, 45*(2), 387–398. doi:10.2307/3069353.

Seo, D., & Desouza, K. (2006). Power-shifting. *Business Strategy Review, 17*(1), 26–31. doi:10.1111/j.0955-6419.2006.00387.x.

Seo, D., & Koek, J. W. (2012). Are Asian countries ready to lead a global ICT standardization? *International Journal of IT Standards and Standardization Research, 10*(2), 29–44. doi:10.4018/jitsr.2012070103.

Seo, D., & Lee, J. (2007). Gaining competitive advantage through value-shifts: A case of the South Korean wireless communications industry. *International Journal of Information Management, 27*(1), 49–56. doi:10.1016/j.ijinfomgt.2006.12.002.

Seo, D., & Mak, K. T. (2010). Using the thread-fabric perspective to analyze industry dynamics: An exploratory investigation of the wireless telecommunications industry. *Communications of the ACM, 53*(1), 121–125. doi:10.1145/1629175.1629205.

Shapiro, S., Richards, B., Rinow, M., & Schoechle, T. (2001). Hybrid standards setting solutions for today's convergent telecommunications market. In *Proceedings: 2nd IEEE Conference on Standardization and Innovation in Information Technology* (pp. 348-351). Boulder, CO: IEEE.

Shell, G. R. (2004). *Make the rules or your rivals will.* New York: Crown Business.

Shetty, R. (2007). The talkative Indian. *Far Eastern Economic Review, 170*(10), 57.

Siemens, G. (1957). *History of the house of Siemens.* Freiburg, Germany: Karl Alber.

Simchi-Levi, D., Kaminsky, P., & Simchi-Levi, E. (2003). *Designing and managing the supply chain: concepts, strategies, and case studies.* New York: McGraw-Hill.

Simon, H. A. (1996). *The sciences of the artificial.* Cambridge, MA: MIT Press.

Sirmon, D. G., Hitt, M. A., & Ireland, R. D. (2007). Managing firm resources in dynamic environments to create value: Looking inside the black box. *Academy of Management Review, 32*(1), 273–292. doi:10.5465/AMR.2007.23466005.

Slater, S. F. (1997). Developing a customer value-based theory of the firm. *Journal of the Academy of Marketing Science, 25*(2), 162–167. doi:10.1007/BF02894352.

Smith, A. C. T., & Graetz, F. (2006). Complexity theory and organizing form dualities. *Management Decision, 44*(7), 851–870. doi:10.1108/00251740610680569.

Stanley, H. E., Amaral, L. A. N., Buldyrev, S. V., Gopikrishnan, P., Plerou, V., & Salinger, M. A. (2002). Self-organized complexity in economics and finance. *Proceeding of the National Academy of Sciences of the United States of America, 99*(1), 2561-2565.

Steinbock, D. (2001). *The Nokia revolution: The story of an extraordinary company that transformed an industry.* New York: AMACOM.

Sullivan, L. (2004, December 13). Team of the year. *InformationWeek.*

Sun-Tzu. (1994). *The art of war* (Sawyer, R. D., Trans.). Boulder, CO: Westview Press.

Tassey, G. (2000). Standardization in technology-based markets. *Research Policy, 29*(4-5), 587–602. doi:10.1016/S0048-7333(99)00091-8.

Thompson, J. D. (1967). *Organizations in action: Social science bases of administrative theory.* New York: McGraw-Hill.

Toutan, M. (1985). CEPT recommendations: CEPT's part in developing a homogeneous, efficient European telecommunications network. *IEEE Communications Magazine, 23*(1), 28–30. doi:10.1109/MCOM.1985.1092415.

Tripathi, M. (2006). Transforming India into a knowledge economy through information communication technologies—Current developments. *The International Information & Library Review, 38*, 139–146. doi:10.1016/j.iilr.2006.06.007.

Tsikriktsis, N., Lanzolla, G., & Frohlich, M. (2004). Adoption of e-processes by service firms: An empirical study of antecedents. *Production and Operations Management, 13*(3), 216–229. doi:10.1111/j.1937-5956.2004.tb00507.x.

Turcotte, D. L., & Rundle, J. B. (2002). Self-organized Complexity in the physical, biological, and social sciences. *Proceeding of the National Academy of Sciences of the United States of America, 99*(1), 2463-2465.

Turcotte, D. L., Malamud, B. D., Guzzetti, F., & Reichenbach, P. (2002). Self-organization, the cascade model, and natural hazards. *Proceeding of the National Academy of Sciences of the United States of America, 99*(1), 2530-2537.

van Wegberg, M. (2004). Standardization process of systems technologies: Creating a balance between competition and cooperation. *Technology Analysis and Strategic Management, 16*(4), 457–478. doi:10.1080/09 53732042000295784.

Verna, A. (2002). *A value network approach for modeling and measuring intangibles*. Retrieved from http://www.vernaallee.com/value_networks/A_ValueNetwork_Approach.pdf

Vojcic, B., Pickholtz, R., & Stojanovic, I. (1991, June). *A comparison of TDMA and CDMA in microcellular radio channels*. Paper presented at the International Conference on Communications. Denver, CO.

Von Neumann, J., & Morgenstern, O. (1953). *Theory of games and economic behavior* (3rd ed.). Princeton, NJ: Princeton University Press.

Walsham, G., & Sahay, S. (1999). GIS for district-level administration in India: Problems and opportunities. *Management Information Systems Quarterly, 23*(1), 39–65. doi:10.2307/249409.

Wernerfelt, B. (1984). A resource-based view of the firm. *Strategic Management Journal, 5*(2), 171–180. doi:10.1002/smj.4250050207.

West, J. (2000). Institutional constraints in the initial deployment of cellular telephone service on three continents. In Jakobs, K. (Ed.), *Information Technology Standards and Standardization: A Global Perspective* (pp. 198–221). Hershey, PA: Idea Group. doi:10.4018/978-1-878289-70-4.ch013.

West, J., & Dedrick, J. (2000). Innovation and control in standards architectures: The rise and fall of Japan's PC-98. *Information Systems Research, 11*(2), 197–216. doi:10.1287/isre.11.2.197.11778.

Westmand, R. (1999). The battle of standards and their road to peace. *On the New World of Communication, 1*(1), 28-28.

Woodruff, R. B. (1997). Customer value: The next source for competitive advantage. *Journal of the Academy of Marketing Science, 25*(2), 139–153. doi:10.1007/BF02894350.

Yan, X. (n.d.). *3G mobile policy: The case of China and Hong Kong, China*. International Telecommunication Union.

Yoffie, D. B. (1996). Competing in the age of digital convergence. *California Management Review, 38*(4), 31–53. doi:10.2307/41165853.

YonHap News. (2005, January 3). GSM royalty, 20% of handset price, need a plan to cope. *YonHap News*.

Yu, J. (2005). The national champion in ICT standard competition: Evidence from an emerging country. In *Proceedings of IEEE Conference*, (pp. 114-120). IEEE.

About the Author

DongBack Seo earned her Doctor of Philosophy and Masters of Science in Management Information Systems from the University of Illinois at Chicago and her Bachelor of Engineering from Hansung University. Prior to pursuing the Ph.D. program, she worked as a software engineer in a wireless communications firm and as a small business owner. Her publications include two books in Korean, as well as a class manual and several chapters. Her papers have been published in many journals (e.g., *Communications of the ACM*, *European Journal of Information Systems*, *Telecommunications Policy*, etc.) and conference proceedings (e.g., International Conference on Information Systems, European Conference on Information Systems, etc.). Her research interests include the areas of adoption of IT/IS-enabled services from the perspectives of individuals and organisations, organisational standards strategy, business convergence, mobile commerce, and analysis of competitive dynamics in rapidly changing industries.

Index